INFORMAL SECTOR
Concept, Dynamics, Linkages and Migration

About the Author

Kishor C. Samal is Professor of Economics at Nabakrushna Choudhury Centre for Development Studies, Bhubaneswar. He has published about 155 research papers in journals of repute. He has to his credit many important books, including Urban Informal Sector (An Exploration of the Informal Sector in a Small City of Orissa); Tax Structure and Budgetary Trends; State, NGOs and Disaster Management; and Poverty, Social Capital and Natural Resource Management. Also, Prof. Samal has completed more than 30 research projects relating to natural resources, poverty, livelihood, displacement and informal sector.

Informal Sector

Concept, Dynamics, Linkages and Migration

Kishor C. Samal

CONCEPT PUBLISHING COMPANY, NEW DELHI-110 059

Cataloging in Publication Data—DK
Courtesy: D.K. Agencies (P) Ltd. <docinfo@dkagencies.com>

Samal, Kishor C. (Kishor Chandra), 1948-
Informal sector : concept, dynamics, linkages and migration / Kishor C. Samal.
p. cm.
With reference to Orissa, India.
Includes bibliographical references (p.)
Includes index.
ISBN 13: 9788180695452
ISBN 10: 818069545

1. Informal sector (Economics)—Government policy—India—Orissa. 2. Rural-urban migration—India—Orissa. 3. Rural development—India—Orissa. I. Title.

DDC 330.954 22

ISBN -13: 978-81-8069-545-2
ISBN-10: 81-8069-545-X

First Published 2008

Published and Printed by
Ashok Kumar Mittal
Concept Publishing Company
A/15-16, Commercial Block, Mohan Garden
New Delhi 110 059 (INDIA)

Phones : 25351460, 25351794
Fax : 91-11-25357103
Email : publishing@conceptpub.com

To my wife

Minati

who sacrificed her research
for my research
activities.

Acknowledgements

There has been an upsurge of debate, discussion and controversy around the concept of informal sector since 1970s particularly in the context of developing countries. The very concept of informal sector assumes a dualism in urban economy. Due to structural adjustment and consequent liberalisation, privatisation and integration of world, the importance of informal sector employment is growing as a safety net and with this trend, the significance of the concept of informal sector, instead of marginalised, is rebounding with vigour in the era of liberalisation. The discussion on informal sector is mainly related to the concept, its linkages with formal sector, rural-urban migration, its growth and dynamics. The present book examines all these aspects by extensively reviewing the literature on it and citing major studies undertaken by the author.

Therefore, the book heavily relies on eight major research projects undertaken by the author, one at G.M. College, Sambalpur and, others at Nabakrushna Choudhury Centre for Development Studies, Bhubaneswar, viz. *Linkages Between Informal and Formal Manufacturing Sectors: Case Study of a Steel City (Rourkela)* and *Dynamics of Informal Manufacturing Sector: Case Study of Sambalpur at Two Points of Time*, both financed by ICSSR; *Migration of Workers to Urban Informal Sector (Case Study of Five Class-I Towns in Orissa)* sponsored by Planning Commission, Government of India; *Linkages of NALCO Angul Sector with the Manufacturing Sector in Orissa* and *Growth of Informal Sector Activities Around NALCO*, both funded by NALCO; *Migration of People to Talcher Coal Mine Regions: A Study of Informal Workers*, financed by MCL; *Liberalisation, Globalisation and Privatisation: Its Impact on Industrial Sectors Including Small-Scale Enterprises and Informal Sector in Russia—Lessons for India*, funded jointly by Institute of Oriental Studies, Russian Academy of Sciences (Moscow) and ICSSR; and *Features and Determinants of Rural Non-Farm Sector in India and Orissa*, an internal project of NCDS.

At the outset, I express my gratitude to Indian Council of Social Science Research (ICSSR), New Delhi, Planning Commission, Government of India, NALCO, MCL and Institute of Oriental Studies of Russian Academy of Science (Moscow) for extending financial support to the studies on which the present book is based.

I am grateful to Prof. T.S. Papola under whose guidance I completed my doctoral dissertation on urban informal sector at Giri Institute of Development Studies, Lucknow in late 1980s and, for that, he was a great guide for me to undertake various research studies relating to informal sector. Prof. R.S. Rao of Sambalpur University created interest in me on informal sector and his inspiration helped me a lot to involve myself on the topic. I am also grateful to him for his constant advice, suggestions and inspiration.

I am grateful to Prof. Kulamoni Samal for his advice for undertaking serious research.

I am also grateful to Baidyanath Misra, N.K. Sahoo, R.K. Meher, Golak Nath, Sakti Padhi, Hutasan Purohit, K.K. Mohanti, N.C. Kar and others for their suggestions at different points of time.

Those who worked in the various projects are Biswabas Patra, Damodar Jena, Shibalal Meher, Pradeep Nath, Sanjukta Pattnaik, Srijit Mishra, P.K. Mishra and others. P. Jena, G. Biswal, C. Mishra, S. Sahu, P. Behera, Sudhir Dora, A. Banerjee, R. Mohanty and many others worked in various projects for field study. My students who were mostly friendly to me and who helped me a lot in these projects are P.L. Das, Dipak, M. Chaini, Rajesh, Durlava, Ashok and Arun Khuntia. All of them deserve my thanks.

I am thankful to Labour Commissioner, Asst. Labour Commissioners, Government of Orissa, officers of various Municipalities, NACs and various other Departments of Government of Orissa and Government of India, executives of MCL, NALCO, RSP, L&T and others who ungrudgingly helped us to get secondary sources of data and various information relating to our studies.

My special thanks are due to Biswabas Patra who helped me in most of the projects, either working in the project or from outside even when not working in the project. Subrata Ray helped me a lot to give finishing touch to the present book.

I express my gratitude to the librarians of GIDS, Lucknow, National Library, Kolkata, RDC (ND) Library, Sambalpur, NCDS and others. My special thanks are due to Dr. R.N. Mohapatra, Librarian of NCDS, who tried his best to help me to be up-to-date on the literature on informal sector.

I express my sincere thanks to all the respondents who provided information through structured questionnaire and group discussions for our various studies.

I will fail in my duty if I don't mention my *adda* at Raj Hotel/Bihari Hotel of Sambalpur, which were treated by us as 'Coffee House' of the intellectuals of Sambalpur, where we continue to discuss on various topics, which in turn, sharpen our capacity on verbal logic and helped me a lot to write the present volume.

My wife Minati's co-operation and daughter Mitra's help are praiseworthy.

Those who helped me in word processing and typing are Md. Ushman, P.K. Mallia, N.K. Mishra, ABC Production (Sambalpur), First Print (Bhubaneswar), Sujata Pani and others also deserve my thanks.

Once again, I thank them all from the core of my heart.

My special thanks are due to the publisher, A.K. Mittal of Concept Publishing Company to bring out this volume.

I hope, the book will be helpful to scholars and students of Economics, Sociology, Anthropology, Geography, Public Administration, Political Science, Urban and Regional Science and, Business Management and the leaders of trade unions, social activists and NGOs working among unorganised workers.

Kishor C. Samal

e-mail: kishorsamal_05@yahoo.co.in

Contents

Abbreviations

AIT	Asian Institute of Technology
BE	Bazar Economy
CPR	Common Property Resource
DIC	District Industries Centre
DME	Directory Manufacturing Establishment
DPA	Directly Productive Activities
ENIS	Entrepreneurs in Informal Sector
EPW	Economic and Political Weekly
ERRP	Economic Rehabilitation of Rural Poor
FS	Formal Sector
FWP	Food for Work Programme
IBRD	International Bank for Reconstruction and Development
ICLS	International Conference of Labour Statisticians
IDS	Institute for Development Studies
ILO	International Labour Organisation
IMF	International Monetary Fund
IMS	Informal Manufacturing Sector
IRDP	Integrated Rural Development Programme
IS	Informal Sector
IWFS	Informal Workers in Formal Sector
IWIS	Informal Workers in Informal Sector
LC	Lower Circuit
MCL	Mahanadi Coalfields Limited
NABARD	National Bank for Agriculture and Rural Development
NALCO	National Aluminium Company Limited
NDME	Non-Directory Manufacturing Establishment
NFM	Non-Farm Migratory
NFP	Non-Farm Proper

NSS	National Sample Survey
OAME	Own Account Manufacturing Enterprises
OECD	Organisation of Economic Cooperation and Development
OSIC	Orissa Small Industries Corporation
PSU	Public Sector Undertaking
RIS	Rural Informal Sector
RMW	Rural Main Worker
RNA	Rural Non-Agriculture
RNFS	Rural Non-Farm Sector
RNFSW	Rural Non-Farm Sector Workers
RUM	Rural-Urban Migrant
SAM	Social Account Matrices
SAP	Structural Adjustment Programme
SCP	Small Capitalist Producer
SEEUY	Self Employment of Educated Unemployed Youth
SIC	Standard Industrial Classification
SNA	System of National Accounts
SOC	Social Overhead Capital
SSI	Small Scale Industries
TRYSEM	Training of Rural Youth for Self Employment
UC	Upper Circuit
UIS	Urban Informal Sector
UNCRD	United Nations Centre for Regional Development
UNO	United Nations Organisation
USSR	Union of Soviet Socialist Republic
WDR	World Development Report

1

Evolution of the Concept of Informal Sector

The demographic feature of the urbanisation process, namely, natural increment of population, rural-urban migration resulted in a substantial increase in urban labour force. However, a large part of the rapidly growing urban labour force remains unemployed and underemployed due to the excessive growth rate of urbanisation over that of both industrial production and urban employment. Under the capitalist mode of production, the absolute increase of capital is not usually accompanied by corresponding rise in general demand for labour. Generally economic development based on large-scale, capital-intensive structure, fails to generate adequate and equitable employment and income opportunities in the modern formal sector. As a result, the surplus labour including most of the rural-pushed migrants, unable to get absorbed in the formal sector, has been forced to find its own source of employment and livelihood in a variety of productive activities in urban centres which constitutes, what is now designated as "informal sector".

The lineage of the formal-informal dichotomy is seen as, on the one hand, in the dualistic model of development, and, on the other, in the paradigm of dependency exploitation (Samal: 1990a). In the former sense, informal sector (IS) is a transient sector, characterised by subsistence principle and self-employment but tending to grow into modern capitalist sector. The dependency model sees the growing inequalities between IS and formal sector (FS) as a by-product of the exploitative framework between the core and the periphery at the international, national, regional and even at the local level. International dualism apparent in North-South divide penetrates into the sub-national level in concert with market mechanisms, existing institutional arrangements and public policy to exacerbate internal dualism between various regions and sectors leading to the comparative attractiveness of industrial over agricultural investment and to the growing economies of agglomeration found in the industrial cities which not only drains labour but also capital, raw materials and other resources from the countryside inhibiting the proper growth of agricultural sector in the rural areas. This tendency along with the prevalent institutional arrangement leading to the ever growing concentration of productive assets in the hands of few dynastic industrial houses produces dualism creating unfavourable terms of trade for agriculture and informal non-farm activities and also obstructs the growth of their productivity. The capital-intensive nature of investment together with imported advanced technology and highly formalised management institutions reinforce the process of unequal internal development, creating what has been termed as FS-IS dichotomy in the urban economy.

The three dualistic relationships, North-South, Urban-Rural and FS-IS, are all related and integrated in a complex manner which differs from country to country depending on various factors such as resource endowment, demographic characteristics, technology and institutional framework and their historical and present conditions. These major strands of thinking and arguments on the genesis and role of IS have been quite well-known.

There has been an upsurge of debate, discussion, dialogue and controversy around the concept of IS since 1970s particularly in the context of developing countries which have experienced a pattern of urbanisation characterised by engagement of large part of labour force in the low productivity low-income activities outside the organised modern sector.

The very concept of informal sector assumes a dualism in urban economy. There has been a long discussion from social dualism to urban dichotomy. Most of the discussion is centred round the concept of formal-informal dichotomy predominantly in terms of contrasting characteristics of economies and participants in two sub-sectors of the urban economies for a proper understanding of the concept of informal sector. It is, therefore, worthwhile to look into the various approaches to dualism and dichotomy advanced by economists and social scientists at different points of time and in different context.

Macro Dualism

Boeke is perhaps the first social scientist to discuss dualism in a society. He defines social dualism as the clashing of an imported social system with an indigenous social system of another type (Boeke: 1953). According to him, a dual society is composed of two segments; namely, traditional and modernised. The traditional society moulded by fatalism and resignation has limited needs with no profit motive.

Boeke's theory of social dualism has been discarded on various grounds, especially because of his wrong assumption that economic boundaries were coincidence with cultural boundaries (Higgins: 1966). Though Higgins does not agree with Boeke's concept treatment, still he recognises dualism as an important characteristic for most of the developed countries. He talks of technological dualism in those countries where there is an industrial sector with relatively fixed capital coefficients along with a rural sector where technical coefficients are variable (Ibid.).

The concept of dualism in macro sense has its origin in the celebrated Lewis model. The model postulates that the capitalist sector expands, when the capitalist surplus is reinvested taking more people out of the subsistence sector into the capitalist employment at constant real wage at subsistence level which is below the marginal productivity of labour (Lewis: 1954). This process will continue until the labour surplus disappears. While discussing this process, Lewis assumes the existence of two sectors in an economy, a capitalist sector which uses reproducible capital and, a subsistence sector[1] which does not use it and where marginal productivity of labour is negligible or zero or even negative. There is an implied assumption of non-existence of urban unemployment in the model.

Lewis model further elaborated by Fei and Ranis (1967) who assume coexistence of two sectors in labour surplus type of underdeveloped economy: subsistence agricultural sector and commercialised industrial sector. However, the macro dualistic models of Lewis and, Fei and Ranis ignore most of the structural conditions in contemporary developing countries. The labour released in the subsistence sector is found to be at higher rate than the capitalist sector can absorb. They also failed to see a segmented urban economy which is empirically recognised. The models offer limited analytical explanation of the problem of employment and rural-urban migration in the developing world.

A dichotomy in an urban economy is recognised by Todaro model. In contrast to Lewis-Ranis-Fei model of one-way migration, the migration is regarded as two-stage phenomenon in Todaro model. In the first stage, an unskilled rural worker migrates to an urban area and initially spends a certain period of time in the so-called urban traditional sector and in the second stage, he eventually attains a job in more permanent modern sector (Todaro: 1969, 1971, 1976). In conventional individualistic approach, the model does not recognise the structural characteristics of the place of origin of the migrants which may compel them to leave it. There is a push-effect—the force of expulsion (S. Amin: 1974). The nature of this push-effect is closely related to the transformation of socio-economic conditions in the rural areas. Moreover, contrary to the model, most of the rural migrants do not graduate from urban traditional sector to modern sector. Despite its limitations, Todaro model is an improvement over all other macro dualistic models because of its capability to recognise a dichotomy in urban economy which is so evident in developing countries.

Urban Dichotomy

There are some social scientists (e.g. Geertz: 1963; McGee: 1975; Santos: 1979) who recognise a dichotomous urban economy in a framework other than formal-informal classification. Santos has considered the city consisting of two sub-systems, the upper circuit and the lower circuit. The upper circuit is direct result of technological progress and the monopolies are its leading representative elements; and the lower circuits of small-scale activities is almost exclusively for the poor (Santos: 1979). The dependency of the lower circuit upon the higher circuit is assumed. Santos has mentioned about two dozen of distinguishing features of the lower circuit[2]. According to him, in the lower circuit, haggling—price negotiation between buyers and sellers—is the rule. This particular feature of haggling over the terms is similarly present in the 'bazar' economy as described by Geertz.

Geertz writes "the bargaining, sliding price system thus tends to focus all the traders' attention on the individual two-person transaction: the aim is always to get as much as possible out of the deal immediately at hand" (Geertz: 1963). In a 'bazar'[3] economy, the total flow of commerce is fragmented into large number of unrelated person-to-person transaction which are mainly *ad hoc;* on the contrary, trade and industry takes place through a set of impersonally defined social institution in the firm-centred economy. The 'bazar' deals with unbulky, easily portable and storable goods.

McGee has drawn extensively on Geertz in describing the economic structure of cities in developing countries. He has used the same dichotomy—bazar-type economy and firm-type economy[4]—on the basis of different pattern of unemployment and underemployment as well as of employment opportunities. In a similar fashion, he has also divided the whole economy into traditional sector and modern capitalist sector. To him, a traditional sector has a labour-absorbing function and consists of an interlocking urban-bazar, rural-peasant agricultural structure in which the former depends to a considerable degree of its food supplies on the latter (McGee: 1975).

The dependency theory underlines the fact that the structure of the Third World societies and their cities was primarily a result of the manner in which they had been integrated into an international capitalist system since the 16th century and, that the continuing pattern of integration prevalent in the Third World since there has been responsibility for a particular socio-economic formation-peripheral capitalism—which is very different from the capitalist mode of production in the developed capitalist countries (McGee: 1982).

McGee has underlined certain features of peripheral capitalism such as: (a) a combination of capitalist and non-capitalist mode of production under the hegemony of the first mode, (b) a combination of these two modes in which one is grafted on to another and is characterised by a series of foreign introduced leaps rather than an entirely domestic evolution (Ibid.). There are three reasons for which dominant mode of production are not acting to dissolve all non-capitalist modes. First, capitalist mode of production enjoys certain advantages by preserving them. Second, non-capitalist mode also yields some benefits to the governments of peripheral areas. Third, people with the non-capital mode are remarkably adept at preserving their activities.

Others (e.g. Reynolds: 1969) have also brought distinction between modern sector and traditional sector in an urban economy. The traditional sector by Reynolds is drawn upon Geertz's 'bazar' economy. But the assumption of the traditional-modern dichotomy that the westernised modern sector is the source of dynamism and change and the traditional sector slowly withers away has been refuted by many scholars (e.g. ILO: 1972; Sethuraman: 1976; Weeks: 1975) on various grounds.

Similarly, the urban dichotomy has been labelled in different terms that are not necessarily identical. They are rich and poor sectors (Weeks: 1977); protected and unprotected sectors (Harberger: 1971). Other terms similar to informal sector have appeared in the literature such as trade-service sector, community of the poor, street economy, unremunerated sector, unstructured sector, irregular sector, casual work, non-plan activities, people's economy, non-westernised sector, urban subsistence sector, etc. (A. Amin: 1996).

Some scholars (e.g. Lewis: 1979; Mazumdar: 1973) have brought dichotomy not in an urban economy as such but in a labour market of an industry as a whole. While studying the labour supply in early industrialisation in case of Bombay textile industry, D. Mazumdar (1973) mentioned that supply of labour to textile industry consisted of two basic types—the stable and unstable from the point of view of commitment to industrial work. He cites the example of *badli* system in the textile industry. The *badlis* were literally substitute labour. They were employed on a day-to-day basis alongside a permanent core of workers paid on a monthly basis. These *badli* workers are unstable labourers. To Mazumdar, *badlis*, were, in some cases, probationaries. The *badli* system was in all probability a system of screening and recruitment of the stable core out of the mass of migrants coming into labour market.

But all these 'permanent' and 'temporary' and 'irregular' workers are clubbed together in one sector (i.e. formal sector) by Nelson (1979). Nelson (1979) has cited the examples with data from different countries including India. To him large units all over the world employ casual workers to meet peak load but in developing countries, formal sector units use them intentionally to avoid wage and labour regulations applicable to permanent workers. Thus, while the employment of temporary workers is a system of screening and recruitment of the permanent and regular workers as found in Mazumdar's study, according to Nelson, it is mainly to avoid minimum wage and other benefits available to regular workers by law. It seems Nelson's view is more appropriate.

In India, the most commonly used classification is one which divides the economy into organised and unorganised (Bhardwaj: 1979; Joshi and Joshi: 1976). Though Joshi and Joshi have used the size of the establishment for division in their study of Bombay, they highlight the distinguishing features of the organised and unorganised sectors under three heads: (a) market structure, (b) technology, and (c) relationship with the government. Though the distinguishing features by Joshi and Joshi (1976) provide a reliable basis for classifying the labour market and workers, there are few limitations in their dichotomy. The designation—'organised' and 'unorganised'—has no special significance as the degree of organisation in the unit is not regarded as a basic distinguishing feature in the classification. Again, since the unorganised sector subsumes considerable heterogeneity, the analytical value of this dichotomy is limited.

Formal-Informal Dichotomy

In view of some limitations of the earlier classification, the use of informal-formal sector dichotomy was used as a conventional method of analysing the structure of urban economies. Informal-formal dichotomy has been put forward to seek an analytical terminology to describe a duality that avoids bias against low income sector inherent in traditional-modern dichotomy.

Hart was first to use informal-formal dichotomy in the study of one migrant group, the Frafas in an urban area of Nima in Acra (Ghana). Though Hart's article was published in 1973 later than the Report of ILO (ILO: 1972), his article is regarded as the first in the sense that he presented the paper at the conference on "Urban Unemployment in Africa" in Institute of Development Studies, University of Sussex in September 1971. Subsequently, the paper was published in *The Journal of Modern African Studies* (Vol. 11, No. 1, 1973) in a revised form under the title 'Informal Income Opportunities and Urban Unemployment in Ghana'.

Though, the term 'informal sector' is used by Hart first, this sector has also existed in the past. The plight of migrant workers—the surplus people in latent form—engaged in begging, cart-pulling, manure collection, etc. in the cities of England during the period of industrial revolution has drawn the attention of the eminent scholars like Marx and Engels (Marx and Engels: 1976). Lewis has also mentioned that "the phenomenon is not, however, by any mean confined to the countryside. Another large sector to which it applies is the whole range of casual jobs. These occupations usually have a multiple of the number they need, each of them earning very small sums from occasional employment" (Lewis: 1954).

Myrdal has clearly mentioned that the category of 'informally' organised or unorganised or very loosely organised in South Asian countries embraces a wide range of heterogeneous activities that have only one attribute in common—a set of institutional properties differing from those observed in either the 'formally' structured westernised units of production or the traditional rural craft (Myrdal: 1968). Some of them are geared to some modern demands. Most have tended to perpetuate traditional pattern such as emphasis on the family as the nucleus of production organisation. Myrdal has given examples of service sectors, crowded field of retail trade including hawkers and peddlers, low-producing field of personal services including porters, rickshaws and taxi drivers and manufacturing activities including traditional urban crafts and small-scale activities.

Thus, there are descriptions of a sector by eminent scholars like Marx, Lewis and Myrdal which are very similar to informal sector but unnoticed by the scholars on the topic of informal sector. For this, Hart is regarded as the first scholar to use the concept of 'informal sector'.

To Hart, the distinction between formal and informal income opportunities is based essentially on that wage-earning and self-employment. The key variable is the degree of rationalisation of work—that is, whether or not labour is recruited on a permanent and regular basis for fixed rewards. Most enterprises, run with some measure of bureaucracy, are amenable to enumeration by surveys and as such constitute the 'modern sector' of the economy. The remainder—that is those who escape enumeration—are variously classified as the 'low producing urban sector', the 'reserve army of underemployed or unemployed', the 'urban traditional sector' and so on. In practice, informal activities encompass a wide-ranging scale, from marginal operations to large enterprises, whether their productivity is relatively high or low remains a question of empirical verifications.

Hart gives a typology of different opportunities in urban areas referring to activities or roles and not persons since actual individuals are often to be found on both sides of the analytical divide and in more than one capacity. His typology is: (1) formal income opportunities which include (a) public sector wages, (b) private sector wages, and (c) transfer payments, (2) informal legitimate income opportunities including (a) primary and secondary activities, (b) tertiary enterprises with relatively large capital inputs, (c) small-scale distribution, (d) other services, e.g. musicians, etc. and (e) private transfer payments, and (3) informal illegitimate income opportunities which includes (a) services (prostitution) and (b) transfers (petty theft like pick pocketing) (Hart: 1973). However, to him, this list is not exhaustive.

The difference between the analysis of Hart and the earlier models of macro dualism is in the identification of 'new income generating activities' in the informal sector. For the first time, the heterogeneity of people who were earlier grouped together as a residual was recognised by his study. In spite of these merits, Hart's dichotomy does not recognise the existence of wage labourers in the IS and has also not specified the criteria by which the self-employed in one sector can be distinguished from those in other sector. The distinction between formal and informal income opportunities based essentially on that between 'wage-earning' and 'self-employed' precludes the possibility of workers being engaged by the IS operations and is not, therefore, an adequate basis for analysing the sector (Davies: 1979). The approach which simply lists the type of job regarded as informal is not sufficient by itself, because of the overlap between categories of IS and FS activities. Each one of these activities is also found in the FS. There is also no clarification of Hart's definition of 'target group' and 'informal-income generating activities'. Certain terms such as informal income generating activities, petty capitalism, urban sub-proletariat, unremunerated sector, unorganised sector and self-employed individuals which has been used interchangeably by Hart lead to more confusion over the terms.

The ILO Mission to Kenya followed the analysis of Hart in utilising the characteristics of enterprises as the basis of dichotomy but it enumerated detailed and specific characteristics of the two sectors to identify the target group more distinctly mainly for the policy measures. The Mission defined that informal activities are the way of doing things, characterised by (a) ease of entry, (b) reliance on indigenous resources, (c) family ownership of enterprises, (d) small-scale of operation, (e) labour-intensive and adapted technology, (f) skill acquired outside the formal sector system, and (g) unregulated and competitive market (ILO: 1972). Informal sector activities are largely ignored, rarely supported, often regulated and sometimes actively discouraged and often, in some cases, harassed by the authorities. This aspect is also mentioned by Santos, while explaining the 'lower circuit' in his dichotomy. The characteristics of IS, mentioned above, are responsible for the low income of the IS participants for which the entrepreneurs in this sector must act continuously to protect himself against the risks. The characteristics of FS activities are the obverse of these features of IS such as (a) difficult entry, (b) frequent reliance on overseas resources, (c) corporate ownership, (d) large-scale of operation, (e) capital-intensive and often imported technology, (f) formally acquired skills often expatriate, and (g) protected markets (through tariffs, quotas and trade licences) (Ibid.).

The ILO Report seems to clarify in more details what some earlier scholars like Santos and McGee had implied in their writings. For instance, Santos' distinction on the basis of relationship to government and the feature of McGee's bazar economy such as family ownership of enterprise has been made explicit by the ILO Mission Report. Again, it is the 'enterprise' and not 'individual' that are classified into two sectors which is to be considered as a positive step in defining informal sector.

After that Report, the formal-informal dichotomy got the official stamp urging a change of government attitude towards the IS participants. But simultaneously it also attracted a volley of criticisms (see for exp., Leys: 1973; Sinclair: 1978). First, the characteristics which was proposed keeping Kenya in view are appropriate to Kenya and do not seem to be relevant in all cases. Second, the Report generally refers to self-employed and the like and largely neglects wage workers while explaining the IS concept. So, while prescribing policies, the implied assumption is the unity of interest inside the IS while there is diversity of interest between the employers and wage labourers within the sector. The IS denotes primarily a system of very intense labour exploitation with very low wages and often very long hour under-priced by the constant pressure for work from the 'reserve army' of job seekers. Third, the activities considered by the Report forms only a small part of the greater system of exploitation on which the FS rests. The various IS activities provide cheap goods and services meant for the poverty life style of those whose work makes possible the very high profits and wages in the FS. Fourth, its faith in a reformed capitalism without contradiction is somewhat illusory. Fifth, the 'informal sector' of the ILO Report is somewhat romanticised.

Sixth, there is a controversy over the issue of entry—the first characteristics of the IS mentioned by the Report. Usually, the notion of 'ease of entry' to the IS is based on various factors such as: (a) requirement of little capital to start, (b) possibility of viability and competitiveness at small size, and (c) importance of personal factor. But the various studies (e.g. Sinclair: 1978) confirm that the entry is not free as is made out to be in the line of the ILO Report. There are constraints on beginning work and subsequently on consolidating their position. The entry constraints are divided into four sets: (a) control of resources, (b) urban environment (i.e. change in labour market condition), (c) government policy towards informal sector, and (d) capitalist penetration into the IS. A distinction can be drawn between conditions of entry to various occupations for long-established town residents as against conditions facing new arrivals. However, migrants to towns do not necessarily find it impossible to acquire jobs and earnings comparable with those held by urban residents.

Seventh, major criticism on the concept of informal sector, sometimes, are grouped under three headings: (a) aggregation, (b) linkages, and (c) dynamics (Sinclair: 1978). By the treatment of its constituents as a 'sector', with the attendant implied shared production function, analysis of each individual part of the whole is impeded. Which parts of the sector may easily be entered is not clear. However, at present, various studies have brought differentiation between different sub-sectors of the informal sector. The relationship of the informal sector with others in the city economy is left unexplored. Whether these relationships are useful linkages or exploitative is essential to decide if policies are to be formulated for this sector. Such *a priori* determination of characteristics of the analytical frame may not be reliable in so far as it tends to ignore the relationship between FS and IS. This aspect has been taken up by various recent studies on informal sector. The dynamics of these operations through time is not made clear by the conventional treatment of the 'informal sector'. The way in which its element may grow or change in connection with which other elements of the city economy is not illuminated.

Eighth, the classification into two categories is crude and simple (Bromley: 1978, 1980, 1981). Ninth, the division of urban economy into FS and IS is logically inconsistent in that it is assumed that several different variables can be used to categorise a given economic activity and yet no multivariate analysis procedure is used in classification (Bromley: 1978). Tenth, there is a mistaken belief that a single policy prescription can be applied to the whole informal sector when there are different categories of activities. Eleventh, there are various other deficiencies of the concept such as : (a) neglect of rural informal sector, (b) lack of clarity about existence of other sector, (c) a tendency to confuse neighbourhood, households, people and activities with enterprises, and (d) absurd consideration of urban IS and urban poor as synonymous (Ibid.).

It may be noted that in spite of the various criticisms, difficulties, confusion and deficiencies in the concept of IS at the initial phase, surprisingly there was a rapid diffusion of the concept since 1971 by the leading institutions such as ILO, UNO, IBRD, UNCRD, IDS and others. More particularly, the concept of IS got an international recognition due to a series of country studies and city studies conducted by ILO under its World Employment Programme. World Employment Programme was launched by ILO in its 1969 session. The country studies were the Colombo Mission (1970), Srilanka Mission (1971), and the Kenya Mission (1972) which were undertaken to evolve 'employment-oriented strategy of development'. In view of the range of theoretical, analytical and developmental significance that is associated with IS paradigm as outlined above, it has generated an enormous body of literature (see, A. Amin: 1996).

The concept of IS first advanced by ILO was immediately taken up in Latin America and Carribean countries where a large body of literature with a variety of approaches—statistical and qualitative, theoretical and programme-oriented—has been generated. Good responses were also forthcoming in Africa beginning with Ley's article. In Asia, a study of IS in Calcutta (Dasgupta: 1973) was the first beginning. Dasgupta's (1973) pioneering contribution includes an estimate of the size of IS, based on a thorough scrutiny of occupational data in Calcutta. The article by Dasgupta (1973), possibly the first work on the IS of an Asian city was followed by a more comprehensive study by A.N. Bose (1978).

Subsequently, there was a rapid growth of literature on IS[5]. The most remarkable aspect of this literature is the spectrum of disciplinary origin, e.g., economics, geography, sociology, architecture, planning, urban and regional science, political science, anthropology, public administration, business management, etc.

Rehabilitation of the Concept

The concept of informal sector which had elements of vagueness and over-generalisation has further been refined by several scholars by adding some more characteristics for distinguishing it from formal sector.

Most scholars make the distinction between IS and FS on the basis of their relation to state. According to Weeks, while there is relatively large-scale operations, capital-intensive techniques, high wage rates and salary levels in FS, the IS is characterised by small-scale operations, labour-intensive techniques, low income levels and indigenous ownership (Weeks: 1975). Weeks introduces a new dimension to the distinction between FS and IS. He gives emphasis on the organisational characteristics of exchange relationship and the position of economic activity *vis-à-vis* the state. However, the similar approach is found in the ILO Mission Report, Santo's,

Joshi and Joshi's dichotomy. Of course, they lack precision, incisiveness and focus as evident in Weeks' approach. From Weeks' emphasis on state favours to FS, it seems that he is proponent of 'Structural Disadvantage' argument according to which the smallness of most of the units is the consequence of the structure of an economy in which various policy measures of the government favours the FS.

In contrast, the 'Inherent Disadvantage' argument believes that the difference between small and large is the result of the competition between efficient and inefficient and the state benefits in terms of capital, finance and foreign exchange are open to the FS as also to the small enterprises including IS units but the latter cannot take advantage of these benefits due to the smallness of their size and resources. Emmerij, for example, is apparently in favour of the 'Inherent Disadvantage' argument. To him, income and technological gap between IS and FS will not narrow over time, the IS is transitory and some of its activities have no future at all (Emmerij: 1974). Emmerij characterises IS by (a) small-scale labour intensive technique, (b) supply according to demand-oriented market of low income people, and (c) exclusion from the official statistical enumeration (Ibid.). But Emmerij's view that IS is transitory is not found to be nearer to reality. Proliferation of IS activities is a direct reflection of changing pattern of urban life and is in response to the need generated by it. Further, an expanding FS activities create more demand for IS.

Lubell, in his study of Calcutta, uses term, 'unorganised' and 'informal' as synonymous terms in the IS as the labour market of the last resort (Lubell: 1973). To him, IS constitutes the residual labour market of the last resort. It includes the unorganised parts of transport, trade and service sectors as well a many of the smaller enterprises registered under the Shops and Commercial Establishment Act and a large number of workshops which are referred in India as the household manufacturing sector (Ibid.). Sethuraman (1976a, 1976b) defines the IS in terms of the source of employment of the urban poor. By and large, most poor people work in the IS but all the participants working in IS are not poor and all the poor do not work in IS. However, the merit of Sethuraman's analysis lies in his exhaustive list of different criteria for identifying IS enterprises in different sub-sectors, viz., manufacturing, construction, transport, trade and services, the central theme being the size criteria[6]. By recognising the different sub-sectors within IS, a number of doubts get cleared.

In a study (Dasgupta: 2003) in Delhi, two distinct categories within IS—informal services and informal manufacturing—are recognised, which exhibit different structural and behavioural characteristics. The heterogeneous segments of employment and variations in earnings among the informal workers are also highlighted by a study using the latest labour force survey data in India (Unni: 2005). The various segments among the IS workers enter this sector for varying reasons, sometimes voluntarily and sometimes involuntarily. There is heterogeneity of regular and casual employees and self-employed workers in terms of their earnings. Self-employed workers are further segregated into those operating with and without capital. The earnings of the former is much higher than the latter (Ibid.).

Subsequently, Sethuraman (1981) modified his definition of concept of IS. Since his suggested definition by a single measure—value added per worker—posed problems relating to empirical evidence and use of labour productivity measures from an operational point of view, he defined IS in a general way. To him, it consists of small-scale units engaged in the production and distribution of goods and services with the primary objective of generating employment and incomes to their participants notwithstanding the constraints on capital, both physical and human and, know-how (Sethuraman: 1981). Thus, while his earlier definition of the concept of IS takes into account the 'individual', the subsequent definition emphasises on 'enterprises'. Sethuraman also brings in mode of production as a basis of distinction between the two sectors. To him, FS consists of activities using modern mode of production and organisation comparable to developed world, hence larger in scale of operation as compared to those of the IS (Ibid.).

It is, however, not clear in what sense the term, mode of production is used and what is meant by modern mode of production. Perhaps, Sethuraman has used the term in simplistic sense, not seriously as mentioned by Marxist scholars. It is to be noted that productive forces and relations of production as a specific historical entity form the mode of production, where the means of production and people with their productive experience and skill are called productive forces and social relation of production or economic relation between people is known as relation of production.

Of course, mode of production is an important element in defining and, distinguishing IS and FS. This is also recognised by Breman (1976, 1977). However, Breman neither defines the concept of IS in clear-cut manner nor deals with the structure of the urban economy as a whole. He mainly deals with labour market in an urban economy. By classifying the urban working population into four broad social classes, viz., labour elite/ labour aristocracy, lumpen proletariat, petit-bourgeoisie and sub-proletariat, he just passes a casual comment by mentioning that the contrast made between the FS and IS parallels that between labour aristocracy and a lumpen proletariat (Breman: 1976). In his terminology, lumpen proletariat are the urban residue with criminal tendencies and declassed and, labour elite are those who carry out work which is highly esteemed in society (Breman: 1977). To him, lumpen proletariat have broken all ties with their original environment and who have nowhere to live and no proper and regular contact with others in their immediate surroundings.

Thus, from Breman's analysis of fragmented labour market, at least, we can infer that IS consists mainly of lumpen proletariat, which is not acceptable. Of course, for his case study in a town of Valsad district of South Gujarat, he gives some outlines of IS workforce such as absence of security and regularity in working conditions and of income derived from work, work carried out to a great extent under terms of 'employment by others' and absence of formalised work relationship[7]. He classified local labour market into two extreme categories: (a) low standing workers and (b) secured and social respectable workers. His main conclusion from his field work is that FS-IS dichotomy is analytically inadequate since some similarities are found when distance between two sectors become closer.

Barbara Harriss (1978) asserting that her distinction between organised and unorganised sectors has a strong resemblance to FS-IS dichotomy, uses this classification in her empirical study in few urban centres in South India to examine the organisation of the labour markets in two sectors and comes to a similar conclusion like that of Breman by casting doubt on the validity of using FS-IS dichotomy. Perhaps, for the overlapping of two sectors in her study, the term 'quasi-formal' is used by her.

Thus, while B. Harriss wants to integrate different types of labour from both the sectors, Breman thinks, there is gradation, and emphasises the fragmented nature of the whole labour market. He suggests that instead of applying the concept FS-IS, one should distinguish in terms of different articulated production relations which can be found within the economic system of Third World countries in varying degree and gradation (Breman: 1976). Thus, though Breman advises to define IS on the basis of mode of production relation, he does not define the concept on this basis or on the basis of mode of production. Throughout his paper, he criticises the concept of 'informal sector' for its seemingly incorrect characteristics but uses the concept throughout. He has analysed more in line with the sociological traditions looking particularly to the labour market only and neglecting the whole structure of the urban economy though he assume unity and totality of the urban production system.

Empirical evidence provided by the study of Ahmedabad by Papola (1980, 1981) points towards a distinction between formal sector and informal sector based on mode of production defined in terms of use of wage labour as one such basis. But he has not used this as the basis of analysis in his study in which he defines IS as a segment of economy having certain characteristics which lead to unfavourable conditions for the growth of enterprise and activities in this segment (Papola: 1980, 1981). Papola listed few most often observed characteristics of IS such as: (a) small size of operations with a cut-off point of size of employment of 10 workers,

(b) informal structure and family ownership, (c) no modern technology, production being labour-intensive predominantly manual, low productive technique of production, (d) lack of access to government favours, (e) competitive and unprotected market, and (f) unprotected labour market.

But he concludes that various characteristics are not necessarily consistent with each other and they do not also provide a clear identification. To him, wage is only variable which has the greater advantage to bring distinction between the sectors, suggesting thereby that the labour market can be demarcated between the FS and IS with greater ease than the economic activities. As his empirical result suggests towards a distinction on the basis of mode of production, on this basis, he includes, in a broad sense, independent workers, casual workers working for households, shops, eating houses and repair units in manufacturing sector in the IS (Ibid.). He thinks a definition of IS on the basis of mode of production also finds empirical validation from his study in terms of formalisation of production process, technology, productivity, surplus generation and labour market process.

It is the work of Davies (1979) that one finds a distinction made between the two sectors exclusively on the basis of mode of production and an examination of how characteristics of IS defined in terms of mode of production arise in his empirical study in Hartley—a small town in Zimbabwe—looking in particular at the part played in the interaction of the mode of production of the IS with the superstructure of the FS. He mentions that in the FS, the mode of production determines superstructure whereas the reverse is true for the IS (Davies: 1979). Further, according to Davies, relations of production in FS are based on a highly developed division of labour, whereas the division of labour in IS, if at all, it exists, is rudimentary and horizontal.

It is to be noted that it is difficult to determine the exact mode of production in a particular sector or sub-sector. Even there have been disagreements among the scholars about the existing mode of production in a sector or in a economy. To some scholars (e.g. Papola: 1981), most of the production organisation in developing countries are not amenable to the classification based on mode of production and most of the producers are not easily classified as either capitalist or wage-earners. But in spite of these difficulties, some scholars (e.g. Gerry: 1978; Lipton: 1984; McGee: 1973; Moser: 1978; Remy: 1982) have tried to define the concept of IS on the basis of mode of production.

Some scholars have tried to define informal sector particularly on the basis of its relationship with formal sector. The central hypothesis of a study in Calcutta (Bose: 1978) is that FS and IS operate in basically two different markets and the interdependence between the two sectors is similar to the relationship between a colonial economy and metropolitan economy. This is particularly in case of modern informal sector. In his study of slum industries in Calcutta (Ibid.), he divides IS into two categories: traditional and modern.

He also mentions two types of characteristics of IS, viz., external characteristics and internal characteristics (Bose: 1980). According to Bose, the basic internal factor which interconnect the various external characteristics and show a special identity to IS is its relative inability to get sufficient amount of resources at relatively cheaper prices and then selling its products. In considering this relative inability, he conceives a hierarchy of informality (or formality)[8]. Hence, one is formal *vis-à-vis* the sector below it and is informal *vis-à-vis* the sector above it. Thus, Bose (1980) explains the concept of IS in a relative term. In theoretical sense, it is more appealing but in practice, it is very difficult to identify one unit in absolute term. But the merit of his analysis lies in the vivid picture of exploitation of IS by FS. According to him, the development of FS is due to underdevelopment and backwardness of IS.

Similar to the conclusion of Bose, the exploitation of IS by FS through sub-contracting and linkages is found in other studies in India (Harriss J. 1982; Papola and Mathur: 1983; Romatet: 1983). Various other studies outside India most of those have brought distinction between the two sectors mainly on the basis of mode of production (Davies: 1979; Godfrey: 1979; Moser: 1978; Remy: 1982) come to a more or less same conclusion in identifying IS.

Size Criteria

In some of the empirical studies, efforts have been made to demarcate informal sector and formal sector. Most of these studies followed one or other criteria such as type of working class (Breman: 1977), types of occupation (Birbeck: 1978; Bromley: 1978; Dhesi and Wadhwa: 1983; Guisenger and Irfan: 1980; Harriss B. : 1978; ILO: 1972; Mazumdar A. : 1980; Papola: 1981; Sethuraman: 1976a, 1976b; Souza and Tokman: 1976; Sreeramamurthy: 1986), income, place of work, etc. or combination of more than one criteria (Romatet: 1983; Sanchez *et al.*: 1981). The size criteria is used most often and except few (Bose: 1978; Harriss B.: 1978; Souza and Tokman : 1976), the cut-off point is ten, that is, ten or less than ten employees including part-time, casual and independent workers in a unit.

Thus, usually the size criteria is used to eliminate enterprises employing more than ten workers from the universe of IS on various assumptions. First, an enterprise normally requires a significant amount of capital and managerial skills once it reaches the employment size of this level. Second, these units are unlikely to have the orientation of an IS enterprise. It looks more appropriate from the context of Indian situation where particularly in manufacturing sector, the registration of enterprise is obligatory under the Factories Act, as the size of employment reaches ten workers with the use of power. But though the size criteria is used in most of the empirical studies, some think that it creates some problems in coordinating the theoretical notion of the concept of IS with its empirical identification on certain economic counts, scale of production and turnover, capital intensity, productivity and surplus generation (Papola: 1981).

Trichotomy

One dimension of the concept is the trichotomy assuming informal sector as one of three segments such as: (a) modern, traditional and informal sector (Lewis: 1979), (b) formal, informal and domestic sectors (Joshi: 1980), (c) modern, intermediate and IS (Steel: 1976, 1977), (d) public sector, organised sector and unorganised sector consisting of (i) quasi-formal and (ii) IS (Sanchez et al.: 1981), and other trichotomy without IS (Gerry: 1979; Harriss, J :1982; Scoville: 1977). While Lewis (1979) trichotomy is in the macro sense seeing the economy as a whole in modification to his earlier position (Lewis: 1954), others gave a trichotomy scheme in an urban economy. Though Aziz (1984) assumes trichotomy in his analysis; in fact, he distinguishes between the two sectors, FS and IS, and divides the latter into two categories. Joshi and Joshi's suggestion to include domestic sector is not practicable.

Sanchez et al. (1981) classification of unorganised private sector into two categories, quasi-formal and informal in their trichotomy scheme is objectionable since the very basis of 'income' for their classification is questionable. Steel (1977) also takes income/wage as one of the main criteria for classification of an urban economy. He has explained characteristics of four urban force categories in a tabular form, viz., (a) modern, (b) intermediate, (c) informal, and (d) unemployed. He assumes that marginal productivity of labour is zero in IS while it is low in the intermediate sector. But this assumption of marginal productivity of labour being zero has been rejected long back.

One study (Harberger: 1971) showing evidence from India concludes that marginal productivity of labour is not zero. Again, Steel's distinction on the basis of technology faces the same problem of clarity. To him, there is small intermediate technology in the intermediate sector and very small and traditional technique in IS but he has not mentioned any criteria to distinguish between the two technologies. Of course, he introduced 'intermediate sector' borrowing the term 'intermediate technology' from Schumacher (1973) who himself has not clearly defined the term. Moreover, the central hypothesis of the trichotomy scheme of Steel (1977) is based on the untenable assumption that IS is unproductive and FS is productive and hence, to him, IS generates employment by sacrificing output. But whether the productivity of IS is high or low is an empirical question. A growing body of evidence (e.g. Guisinger and Mohammad: 1980) shows that IS is equally productive in comparison to FS. Even in some cases, the productivity and average earnings of IS workers are higher than those of unskilled workers in FS (Waldorfs and Waldorfs: 1983).

Like Steel (1977) and Sanchez et al. (1981), the basis of income and labour productivity is used by others (Bose: 1980; Emmerij: 1974; ILO: 1972; Sethuraman: 1981; Souza and Tokman: 1976; Weeks: 1975) but to bring out a dichotomy in an urban economy. To them, there is low productivity and low income in IS and high productivity and high income in FS. But, though the average income of FS is higher than that of IS, there is a great deal of overlap between incomes in the two sectors (Bromley: 1978b; King: 1974, 1975; Nelson J.M.: 1979).

Significance and Approaches

Thus, the distinction between formal sector and informal sector is not analytically clear. In spite of various difficulties, the significance of the concept of IS is recognised for various reasons (Samal: 1986, 1990a). First, there is importance of urban IS in development effort for generation of income and employment and efficient allocation of resources since the benefit of general development process hardly trickle down to the poorest section of population, particularly in developing countries. So, the concept of IS is able to identify a particular sector including enterprise and persons in urban area which are overlooked either in planning or in integrated development programme. Second, the evolution of the concept of IS as a heuristic device in development studies was a useful antidote to the overly rigid application of unemployment-based studies. The concept has enlarged the narrow notion of unemployment in the development literature.

Third, despite rapid industrial development, the urban economy in most developing countries continue to have large part of their activities in IS. The employment in IS is estimated to be varying from 25 per cent to 70 per cent in the urban areas of many developing countries. The IS also contributes substantially to urban income. Finally, besides generating employment, income and efficient allocation of resources, IS has importance for its role: (a) in use of low amount of capital, (b) the use of local resources and recycling of waste and (c) in providing goods and services which would otherwise be available only at a higher price and in some cases would not be available at all were it not for IS.

The main issues in the discussion of IS-FS dichotomy are that while discussing the dichotomy in an urban economy, the emphasis is given on one part—that is, IS (Ibid.). Second, it is now recognised that IS is a permanent phenomena, not transitory. Third, the discussion also centred round the exchange relationship and interdependence between the two sectors.

We observe that there are various approaches to conceptualise IS:

(a) A simple approach to define IS is to examine the list of productive sector of a country's Standard Industrial Classification (SIC) and assign each to either FS or IS.
(b) Another approach is to characterise IS in terms of economic features such as ease of entry, etc.
(c) IS is also conceptualised on the basis of itinerant workers without the fixed workplace.
(d) Most of the operational definition of IS adopted in empirical research are based on either worker's characteristics or establishment size or mixture of both. There is several definitions of IS in terms of workers characteristics. An earning criteria is used to define IS in terms of workers who earn less than the legal minimum wage. On the other hand, using the workers' characteristics, the concept of IS is based on a distinction between wage and non-wage employment (Richardson: 1984).
(e) The more appropriate to identify IS is the establishment size approach, though there has been some disagreement over the cut-off point, about whether it should be less than five workers, less than ten or even less than twenty-five.

Yet the establishment size approach is more convenient since the empirical identification on the basis of size is likely to ensure at least the basic implicit attributes of IS. However, in 1970s and 1980s, there was no satisfactory method of defining IS either analytically or empirically. Therefore, some analysts, in the past, have rejected FS-IS dichotomy. An alternative to outright rejection of the dichotomy is suggested to disaggregate IS itself (Ibid.).

Refinement of the Concept in 1990s (ILO, etc.)

Gradually there is emergence of clarity on definition of IS. In order to avoid confusion, it is suggested that:

(a) It should be restricted to non-agricultural activities.
(b) It should not include criminal activities.
(c) It should exclude the so-called underground economy.

Setting aside these activities, the application of the IS paradigm as an analytical tool can occur with: (a) a people focus, (b) an activity focus or (c) a habitat focus, depending on which particular aspect of the larger world of the IS one is concerned (Amin A.: 1996).

In a people focus, certain labour and employment characteristics can be used to identify IS. They are: (a) official protection/recognition, (b) coverage of wage legislation and other social security system, (c) predominance of own-account work, (d) absence of trade union organisation, (e) low income and wage, (f) little job security, and (f) no fringe benefits from institutional sources. In an economic activity focus, IS is distinguishable from the rest of the urban economy according to enterprise operation characteristics such as: (a) a very small-scale operation, (b) its ease of entry, (c) unregulated and competitive market, (d) reliance on locally available resources, (e) family ownership, (f) labour intensive and adapted technology, and (g) absence of access to institutional sources of credit and similar support or protection.

In a habitat focus, informal settlements are distinguishable according to land and housing characteristics such as:

(a) unauthorised use of vacant public land, (b) illegal sub-division/rental of land, (c) unauthorised construction, (d) reliance on cheap and locally available scrap construction materials, (e) reliance on family labour and artisanal construction technique, (f) absence of safety standards and regulation, and (g) non-availability of mortgage or any other subsidy.

Though A. Amin (1996) has given exhaustive features in three aspects, it seems that there is nothing new in it.

The Fifteenth International Conference of Labour Statisticians (ICLS) adopted a 'Resolution Concerning Statistics of Employment in the Informal Sector' in January 1993 which was the outcome of a series of activities by ILO Bureau of Statistics during the late 1980s. The resolution on IS adopted by Fifteenth ICLS represents the first, so far only internationally approved technical guidelines for the development of statistics on IS specifying the criteria of an operational statistical definition of IS. The UN Economic and Social Council adopted the revised System of National Accounts (SNA: 1993) in July 1993 upon the recommendation of its Statistical Commission. The definition of IS adopted by the Fifteenth ICLS forms part of the SNA 1993.

A report prepared by the International Labour Office (ILO: 1992) for Fifteenth ICLS identified and compared alternative concepts and definition of IS and noted their advantages and disadvantages. After reviewing the ILO background report, the 15th ICLS undertook to clarify the concept of IS by specifying what is considered to be principal and distinctive features of IS enterprises.

The full description of the concept of IS adopted by the 15th ICLS is as follows: "The IS may be broadly characterised as consisting of units engaged in production of goods or services with the primary objective of generating employment and income to the persons concerned. These units typically operate at a low level organisation with little or no division between labour and capital as factors of production and on a small scale. Labour relations—whether they exist—are based mostly on casual employment, kinship or personal and social relations rather than contractual arrangements with formal guarantees. Production units of the IS have the characteristic features of household enterprises. The fixed and other assets used do not belong to production units as such but to their owners. The units as such can not engage in transaction or enter into contracts with other units, nor incur liabilities on their own behalf. The owners have to raise the necessary finance at their own risk and are personally liable, without limit, for any debts or obligations incurred in the production process. Expenditure for production is often indistinguishable from household expenditure. Similarly, capital goods such as building or vehicles may be used indistinguishably for business and household purposes. Activities performed by production units of the IS are not necessarily performed with the deliberate intention of evading the payment of taxes or social security contribution or infringing labour or other legislation or administrative provisions. Accordingly, the concept of IS activities should be distinguished from the concept of hidden or underground economy."

'Operational definition' of IS is developed from an 'explicit' definition of IS which specifies the criteria to be used to distinguish 'informal' enterprises from 'formal' enterprises. Using SNA terminology, the 15th ICLS brought a distinction between 'household enterprises' on the one hand, and 'corporation' and 'quasi-corporation' on the other hand, and stated that only 'household enterprises' could qualify as IS enterprises[9]. A household enterprise is also known as unincorporated enterprise owned and operated by households or individual household members or by several members of the same household as well as unincorporated partnership and cooperatives formed by members of different households. Household enterprises are distinguished from corporations and quasi-corporations on the basis of their legal status and the type of accounts they held. Household enterprises are production units engaged in the production of goods and services which are not constituted as separate legal entities independently of household or of household members that own them. No complete sets of accounts (including balance sheets of assets and liabilities) for these enterprises are available which could permit a clear distinction between production activities of the enterprises and other activities of their owners and the identification of any flows of income and capital between the enterprises and the owners.

There are two categories of household enterprises: (a) enterprises of employers and (b) own-account enterprises. The former employs one or more paid employees on a continuous basis (as contrast with the employment of employees on an occasional basis and of contributing family workers) while the latter do not.

(a) *Own-account Enterprises*: First categories are own-account enterprises (or informal self-owned enterprises) which may employ family workers and employees on an occasional basis. For operational purposes and depending on national circumstances, this segment comprises either all own-account enterprises or only those which are not registered under specific forms of national legislation (such as factories laws or regulation established by national legislative bodies) is considered as informal.

(b) *Enterprises of Informal Employers*: Enterprises of informal employers are defined in terms of one or more of the following three criteria:

(i) the size of the establishment below a specified level of employment; that is, small size of the unit in terms of employment,

(ii) non-registration of the enterprises, and

(iii) non-registration of its employees (in terms of absence of employment contracts which commit the employer to pay the relevant taxes and social security contributions on behalf of the employment relationship subject to standard labour legislation).[10]

Although these criteria are fairly complex, they do help in delinking IS in a clear and logical way. This is specially so because, in practice, the criteria of (i) legal organisation of the enterprise, (ii) type of accounts kept, (iii) size, and (iv) registration overlap to a greater or lesser extent (Government of Pakistan: 1994). Moreover, they provide a broad idea for constructing a fully operational definition of IS.

From the IS enterprises, the following are excluded: (a) corporation and quasi-corporations, (b) enterprises engaged in agriculture, (c) enterprises engaged in non-market activities, (d) certain own-account enterprises, and (e) certain enterprises of employers. The criteria utilised for identifying IS enterprises can be summarised as follows: An enterprise is classified as informal if it meets any (or all) of the three criteria:

(a) the enterprise or any of its establishment employ less than 10 persons, or
(b) the enterprise is not registered,
(c) it does not maintain a complete account.

The value of this definition resides in the fact that it leans on existing practices for estimating informal employment at a national or macroeconomic level (Charms: 1996).

However, it is criticised that, in the Indian context, it is not desirable to link IS with SNA 1993 (Kumar and Sharma: 2003). It is also not appropriate to follow the operational definition suggested by ILO, ignoring the ground realities prevailing in India. Therefore, it is suggested that the only feasible option is to adopt the worker approach, since it is not feasible to capture IS through enterprise approach by adopting size criterion and/or ownership pattern (Ibid.).

The main features of IS definition adopted by the 15th ICLS can be summarised (Charms: 1996; Hussmanns: 1995). First, IS is defined in terms of characteristics of production units (enterprises) in which the activities take place. Consequently, IS includes all persons who are engaged in at least one production unit of IS during a given reference period, irrespective of their status of their employment whether it is their main or secondary job. Persons employed outside the IS enterprises are excluded whatever may be their status of their employment.

Second, IS is considered as a subset of 'household enterprises' which are classified into: (a) enterprises of employers and (b) own-account enterprises on the basis of employment of paid employees on a continuous basis. Third, IS activities are undertaken to earn one's living or obtain an additional income. Therefore, household enterprises exclusively engaged in production for own final consumption or own fixed capital formation like building of own houses, etc. are excluded from IS.

Fourth, agricultural activities are excluded from IS. Fifth, IS includes establishments located in both urban and rural areas. Sixth, enterprises engaged in professional/technical services rendered by self-employed, doctors, lawyers, accountants, architects, engineers, etc. are included in IS. Seventh, out-workers are included in IS if they constitute enterprises on their own as self-employed persons, provided that they satisfy the criteria of IS definition. Eighth, the issue of whether or not domestic workers engaged in households (e.g. maids, watchman, drivers, etc.) should be included in IS is left open to determination depending upon national circumstances. Ninth, there is no FS/IS dichotomy. Activities which are not informal are not necessarily formal and may be identified as separate categories.

Thus IS comprises a fairly heterogenous set of units which need to be classified by various characteristics such as urban *versus* rural location, kind of activity, type of workplace, size, composition of the workforce, types of ownership and relation with other enterprises. These types of classification may be required in order to show the structure of IS and to identify more homogenous groups for data analysis as target for policies and support programme. For particular analytical purposes, it may be necessary to develop more specific definition of IS by introducing further criteria (Hussmanns: 1995).

We observe that there are various generalisation of IS in recent times. It is an umbrella concept describing varieties of activities producing goods and services through which individuals derive employment and income (Ibid.). These activities range from shoe-shining and hawking to manufacturing and modern workshop.

In most of the cases, IS establishments are registered and operating on a very small-scale. Usually they are not recognised, supported or regulated by the government. They are often compelled by circumstances to operate outside the framework of the law. Sometimes, some IS units prefer to remain unregistered and un-licensed in order to avoid compliance with some regulations in order to reduce production cost. However, IS activities are not necessarily performed with the deliberate intention of evading the payment of tax or infringing some government regulations or labour legislations. In fact, they are unable to offer to comply with existing regulations because certain rules and regulations are irrelevant to their needs and conditions. In this sense, they are not 'illegal' and must be distinguished from 'underground economy'. The IS establishments have little access to the facilities offered by public institutional system. They are mostly beyond the purview of labour legislation. There is usually no enforceable contract between employer and employees, if any. The IS establishments are mostly operated by single individual, although some engage a few paid workers or apprentices. The primary objective to undertake IS activities is to earn a living rather than to maximise profit or to get return on investment.

The numerous studies on IS by individual scholars and various institutions including ILO have made it possible to enumerate the main characteristics of IS units, the conditions under which they operate and their linkages with the rest of the economy as well as the characteristics of the unit's head, his/her households and the labour force employed (Charms: 1996). The main features emerging from more than two decades of IS surveys are:

(a) ease of entry,
(b) small-scale activity,
(c) self-employment with a high proportion of family workers and apprentices,
(d) little capital and equipment,
(e) labour-intensive technologies,
(f) low skills,

(g) low level of organisation with no access to organised markets, formal credit, education and training of services and amenities,

(h) cheap provision of goods and services otherwise unavailable,

(i) low productivity and incomes according to some analysts or, on the contrary, incomes that are notably higher than in the public sector, especially recently in the context of Structural Adjustment Programmes (Ibid.).

However, studies on IS in recent years have come out with conclusions that are often self-evident and repetitive. The results are quantitative without being analytic and have not been of much help in understanding the pattern or nature of interdependencies or the dynamic of the sector (Kundu: 1998).

Our Original Concept and its Modification

In our earlier studies (Samal: 1986, 1990a, 1990b, 1990c, 1990d), following the size criteria, we started defining informal sector units as consisting of shops and establishments employing 10 or less than 10 workers. The first order condition for a unit to be in IS must satisfy the size criteria with a cut-off point of 10 workers including owner-operator, paid and unpaid family workers and hired workers.

The second-order condition is the observance of some features of the unit. The characteristics which are obviously implied from the size criteria are: (a) small-scale operation, (b) direct participation of the owner of the unit either in the manual or managerial work. The other features are: (c) the unit in general has informal relationship with the employees and customers, (d) the institutional factors have no role in determination of wage, the wage being determined on the basis of bargaining between the employer and employees, and (e) there is no security and certainty in employment and income. In that earlier studies, we also recognised the differentiation within IS and hence suggested different sub-sectors. But, empirical identification on the basis of the size as the first-order condition only partially satisfied the concept of IS as distinct from FS in all its characteristics identified as second-order conditions.

So in our follow-on-study (Samal: 1994a, 1997b), while we revisited the same sample after nine years but only for manufacturing sub-sector of IS in the same town of Sambalpur (Orissa), we followed the Labour Composition Establishment Approach since we classified the establishments of informal manufacturing sector (IMS) on the basis of composition of labour, like own-account workers, family labour (both paid and unpaid) and hired workers. Under this approach, we grouped the IMS into three categories as: (a) micro-business, (b) petty producers, and (c) small capitalist producers, besides the original classification of IMS on the basis of inputs/raw materials used. From the findings and analysis of the characteristics and performance of IMS with its classification both on the basis of: (a) labour composition and (b) raw materials used during two points of time, 1985 and 1994, it is observed by our study (Ibid.) that the following categories of establishments should not be included in IMS irrespective of their size:

(a) government sector establishments such as PSU, state undertakings, cooperatives and the like,
(b) units which depend 'fully' and 'wholly' on FS for their inputs and outputs and whose entrepreneurs are nothing but 'out-workers' of the FS establishments, and
(c) units working in 'protective economic environment' that is, units in government sponsored industrial estates, export processing zones, special economic zones, and the like.

The units whose entrepreneurs are 'out-workers' of a FS firm and units starting and working in a protected economic environment, that is, categories (a) and (b) may be distinguished as quasi-formal sector units. In this study (Ibid.), initially, we started defining petty producers as one who works along with paid or unpaid family labourers, the total number of participants being 10 or less than 10, without having any wage worker. Since there was no presence of any wage worker, we came to the conclusion that there is no need of any limit to the size of employment of petty production unit. So we suggested to include all micro-business and petty producers in IMS irrespective of their size of employment. Thus we define as follows:

(1) *Micro-business*: Micro-business is one-man enterprise managed by only one active participant. The unit may be in the name of one or more than one person.
(2) *Petty Producers*: Petty producer is the entrepreneur of the unit who works only with 'paid' or 'unpaid' family labourers of any number. There is no limit to the size of employment.
(3) *Small Capitalist Producers*: In this follow-on-study (Ibid.), in the beginning, we started defining small capitalist producer (SCP) as those who employ hired/wage workers with or without paid or unpaid family labourers, total employees being 10 or less than 10. Since, in more than eighty per cent of SCPs, the number of family labourers (F) was less than the wage workers (W), our plan to eliminate the SCPs (where $F<W$) from IMS did not sound relevant. The size of 10 is also irrelevant since some of the SCPs are working in more informal manner, though the number of employees is more than 10. So, we have introduced some additional features of SCP to include them in IMS without having any size criteria or the criteria of ratio of wage workers to family workers. Thus, the following SCPs may be included in IMS.

(a) *SCPs with 10 or less than 10 workers:* SCPs who employ wage workers/hired workers with or without family labourers, total employees being 10 or less than 10.
(b) *SCPs with more than 10 workers*: SCPs who employ wage workers with or without family labourers, total employees being more than 10 but satisfying the following characteristics: (i) 'unprotected' wage work and (ii) 'unprotected'economic environment.

(i) *'Unprotected' Wage Work*: There is no written or oral contract between the employer and workers which is effective in the court of law or judiciary. The wage work is unprotected irrespective of whether workers are regular, irregular, short-term or long-term. There is neither effective trade union of workers nor any effective government assistance to employees.

(ii) *'Unprotected' Economic Environment:* Units are set up with entrepreneurs' own funds and effort and without any major assistance from public institutional system and government agencies. That is, units have been working in an 'unprotected' economic environment.

Informal Sector in the Era of Liberalisation

There is an increasing tendency of growth of the size of IS in developing countries. The labour released from the subsistence sector and latent surplus population from rural areas are the major factors contributing to the growth of urban population in developing countries and the urbanisation in most developing countries at the present stage of their development is likely to continue to be characterised by a large IS. The IS cannot be seen as transient or temporary nor it is holding ground for migrants but a lasting features in the process of economic development.

While commenting on the likely trends and growth of IS, one must recognise the segmentation within IS. In the development process, the sub-sectors of IS which are linked to industrial development will grow at an increasing rate and other which face stiff competition from FS will grow at decreasing rate. Though there may be competition between FS and IS in certain activities like manufacturing, one sub-sector in FS may be complementary to a different sub-sector in IS. For instance, manufacturing in FS and trading in IS are complementary. In certain other activities, IS is able to continue in its operation for its personalised services in spite of stiff competition from FS. In most other activities, it is more likely that the growth of FS and IS get linked together (Samal: 1990a).

Due to structural adjustment and liberalisation, the growth of IS is likely to be more. In many transition economies (that is, former socialist countries), there are increase in unemployment and reduction in income as workers formerly employed by the state sector face wage cut, loss of social security benefits, fewer working hours, non-payment of wages and/or unpaid leave (*IMF Survey*: 1995a). These workers are joining IS.

The same trend is also observed in developing countries due to liberalisation and structural adjustment programmes. As the FS shrinks due to its lack of competitiveness globally and lack of productivity nationally, many people resort to undertaking their socio-economic transaction and activities in the IS in the Eastern and Southern Africa sub-region. Particularly, women's production activities in IS are affected by competition from cheaper imports following the implementation of macro-economic reforms in the sub-region (Akinboade: 2005). Women have engaged themselves in urban IS activities in order to alleviate poverty among themselves in Tanzania (Ngware and Ngware: 2004).

On the other hand, the process of structural reform in Peru, like several other Latin American countries since the early 1990s, may be the main factor in the relative reduction of the size of the IS, since it has increased the benefits of becoming part of the FS and, resulted in an increase in the cost of becoming and staying in the IS in the urban areas of Peru (Saavedra and Chong: 1999).

In 1993, around two billion lived in developing countries. About one-third of the working-age population in these countries is unemployed and around 85 per cent of them earn a living in the informal economy (*IMF Survey*: 1995b).

Some changes are also expected in labour market due to structural reform and liberalisation measures initiated in India since 1991. Theoretically, it is argued that liberalisation may not lead to much job creation but would instead eliminate the inefficient jobs and would raise the wages of IS workers while lowering it for formal sector employees, thus reducing inequality among wage earners (Roemer: 1986). In India, due to liberalisation, the growth of employment in public sector is likely to decelerate. It is also expected that the introduction of exit policy and the removal of other labour market regulation (leading to labour market flexibility) may shift the labour from FS to IS which, in consequence, may very likely lead to the growth of IS activities.

In fact, due to structural reform in India, there are also some changes. Structural reform measures aim at restructuring the public enterprises and promote employment in the organised private industry by reducing the unit cost of labour, stimulating productive growth or by removing other labour market distortations. Due to liberalisation and structural reforms, several trends are observed in the organised sector in India such as: (a) sub-contracting, (b) casualisation of workforce, and (c) appointment through contractors (Mitra: 1997). All these tend to help in the growth of IS due to shift of activities from FS to IS.

The public sector enterprises are increasingly relying on the small-scale and ancillary units since 1991. Private sector industries in India have substantial vender development programmes related to both captive and sub-contracting units in SSI sector. Many large industries sourcing parts and components from the small industries are manufacturing end-products which are marketed under the brand names of large units. Secondly, during 1980s and early 1990s, the work participation rate by usual status declined while that by current status increased. That is, there has been a slower growth in jobs available on a regular basis in the formal organised sector and increase in time or casual work in IS activities. Finally, appointment through contractors is being increasingly practiced in the organised sector as it assumes the constant supply of labour to the employer (Ibid.).

Due to deceleration in formal sector, particularly public sector employment due to modernisation and privatisation, there may be deceleration in the growth of some sub-sectors of IS particularly trade and service sub-sector around the PSU which are catering to the demand for consumer goods and services emanating from the income of the public sector employees. Similarly, due to liberalisation, there is advent of neo-colonialistic trend through the entry of multinationals and liberalisation of international trade. As a result, we observe: (a) availability of cheap, similar and substitute manufactured goods from foreign countries, and (b) change in taste and fashion of the people due to more exposure as well as aggressive advertisement by multinationals and large firms in print and electronic media. These trends adversely affect the growth of IS particularly in the sub-sector of traditional manufacturing component of IS in both rural and urban areas.

For these trends, there has been all-round decline in cottage industries in Orissa (India) since 1990-91. The number of cottage industries units, investment and employment generation has been gradually declining. The employment generation declined from 1,28,218 in 1990-91 to 39,528 in 2002-03 during liberalisation (Samal: 2000).

But the picture is totally different for modern manufacturing sub-sector of IS. It is now thriving in many third world cities as a result of market forces which are being stimulated through measures like Structural Adjustment Programme (SAP), trade liberalisation, deregulation and privatisation. All these measures have created a new demand for IS activities in the form of proliferation of sub-contracted and piece-rate work and a variety of service sector work generated by activities centring on export, imports and tourism (A. Amin: 1996). In this scenario, many stable and secure jobs in FS are disappearing while jobs are being created in IS.

The global and Indian experience, during the era of privatisation, liberalisation and globalisation, so far has produced the negative impact more often than the positive impact on labour. Unemployment has increased in most countries and regions and, employment has mostly increased in IS and in its more flexible forms—casual, contract and temporary (Papola: 2004). Workers displaced from jobs have mostly found alternative employment in IS. Most new workers have also to resort to IS for earning their livelihoods either as self-employed or wage-earners. As a result, IS, which constituted the major part of the economies of developing countries has increased with globalisation and liberalisation. An increase in the share of IS, implies a deterioration in the quality of employment in aggregate (Ibid.).

Thus, while the concept of IS is traceable into the context of the development strategies of the 1960s and 1970s, its relevance in the 1980s and more particularly in the 1990s and 2000s in India appear in a totally different context. There was resurgence of market economy in the 1980s in most developing countries in Latin America and Africa and later in 1990s in the former socialist countries including USSR as well as in India. The IS began to flourish almost immediately particularly in the former socialist economies as well as in the deregulated liberalised, privatised market economies of the typical developing countries such as India.

Along with this transition, there is also profound change in development paradigm. These changes in development paradigms in the 1990s have greatly increased the significance and importance of the concept of IS. The opening up of the economy and reliance on the market forces in developing countries have led to a greater reliance on IS as a main source of employment. The growing importance of the concept of IS in the context of new development paradigm is reflected in the important Report and Studies like *Report on World Employment, 1995* (ILO: 1995), *World Development Report,1995* (World Bank: 1995). While 1995 ILO Report illustrates the role of IS in reducing underemployment and poverty, *World Development Report 1995* shifts its focus from FS to IS employment.

Thus structural adjustment and transition from central planning often bring wrenching changes for workers and their families. Due to structural adjustment and consequent liberalisation and privatisation, when the economy declines, labour is likely to bear the brunt of the shock while capital flees as labour is less internationally mobile than capital. Shifts from FS to IS employment, like increase in unemployment rate are important features of labour market adjustment, especially in countries with no safety net or unemployment insurance. The IS employment grows during the period of adjustment, as workers laid off in FS seek new job and other household members including women and children join outside IS employment to help offset the fall in household income. This 'informalisation' of employment has characterised both former centrally planned economies/transition economies where IS was underdeveloped and developing countries where IS was already highly developed (World Bank: 1995). Casualisation of labour force, feminisation of jobs at lower end, exploitative working conditions, etc. emerge as inevitable consequences of the ongoing process of economic liberalisation in India. So, there is a strong need for an active and integrated approach towards IS and not to accord residual treatment to it (Kundu and Sharma: 2001).

Thus, due to structural adjustment and consequent liberalisation, privatisation and integration of world, the importance of IS employment is growing as a safety net and with this trend, the significance of the concept of IS, instead of marginalised, is rebounding with vigour in the era of liberalisation.

Notes

1. Lewis (1979) categorically asserts that he talks of subsistence sector, not agricultural sector nor rural sector as assigned by some scholars (e.g. Bhardwaj: 1979; Breman: 1976). Similarly the other sector is capitalist sector, not urban sector. It is to be pointed out that it is not Lewis who first talk of two sectors in an economy as generally assumed. The two sectors such as agricultural and industrial sectors have been recognised since Adam Smith.
2. The distinguished characteristics of the two circuits are explained in details by Santos as follows: (a) The upper circuit (UC) uses an imitative, imported, high level and capital-intensive technology while the lower circuit (LC) uses indigenous, locally adapted and labour-intensive technology having considerable innovative potential. (b) While activities in the UC rely on bank credit, that of LC is based largely on direct and personal type of credit and cash. (c) While the merchandise dealt by UC are of large volume and of high quality, that of LC is small quantity and poor quality. (d) In LC, capital is limited, while it is abundant in UC. (e) The organisation in UC is bureaucratic while in LC it is primitive. (f) Labour is limited in UC while it is abundant in LC. (g) Wages are the dominant form of labour relation in UC but employment in LC is seldom permanent and its remuneration is offered at or below the subsistence level. (h) The UC employs many foreigners while jobs are allocated locally in the LC, occasionally, however, other nationals such as Italians in East Africa work in this circuit

as independent traders. (i) Prices are generally fixed in the UC. In the LC, haggling (negotiation between buyers and sellers) is the rule. (j) While there is long-run consideration in the UC, the short-run consideration is permanent in the LC. (k) In the UC, the goal is to accumulate sufficient capital to permit a reproduction of economic activity. In the LC, capital accumulation is of little, if any. (l) The volume of profits in the UC is high whilst its profit margin are relatively low. In the LC the reverse is true. (m) Advertisement is necessary in the UC. In the LC, advertisement is neither necessary nor possible. (n) Fixed cost is substantial in UC, negligible in LC. (o) Re-use of goods is non-existent in UC while it is frequent in LC. (p) While UC benefits from government assistance, LC is harassed by government. (q) UC depends largely on overhead capital which is not necessary for LC. (r) While UC manipulate demands and create new tastes, LC is created by demand and is highly dependent upon consumption. (s) UC activity tends to dominate the entire economy. The LC, however, tends to be controlled, subordinated and dependent. (t) The LC is locally integrated, while in the UC, local activities are integrated with those of another high-order city within the country itself or abroad. Thus, the 'lower circuit' of Santos is very much similar to the concept of 'informal sector'.

3. 'Bazar' or 'pasar' or traditional market are used interchangeably. By 'pasar', we mean—— "the whole of small-scale peddling and processing activity——. The market place is the climax of this pattern, the focus and center but it is not the whole of it" (Geertz: 1963, p. 30). The features of 'Bazar' economy are elaborately mentioned by Geertz in Chapter 4 of his book (Geertz: 1963, p. 30). They are: (a) The bazar economy (BE) breeds a man fast on his feet. (b) A BE provides a preserve for the exercise of economic rationality independently of non-economic constraints. (c) The BE necessarily develops a specific ethics of its own, for no ordered social activity can proceeds without normative control. To understand the 'pasar' in this broader sense, one needs to look at it from three points of view: (a) the flow of goods and services, (b) regulatory mechanism, (c) social and cultural characteristics. Thus the 'pasar' described by Geertz (1963) resembles to some extent the 'haat' system in India.
4. Instead of using the term 'farm-type economy' McGee (1975) used the term 'capital-intensive sector'. To him, capitalist-intensive sector is one "—where productivity is high, the possibilities for employment are limited by labour-destroying innovations. On the other hand, in the 'bazar-type economy' the possibilities for employment are much greater even though returns are much smaller and the end product of this situation is frequently—a

condition of 'shared poverty'—— and less satisfactorily, subsistence urbanisation" (McGee: 1975, p. 69)—"broadly speaking—there are different pattern of unemployment and underemployment as well as of employment operating in the two economic sectors" (p. 72).

The bazar sector can absorb increasing labour because of its following features: (a) The institutional basis of the enterprise is the family. (b) Its system of flow of goods and services allow introduction of more labour. (c) It has something of a self-inflationary quality. (d) The population movement and the flow of goods between it and the peasant economy are common. (e) The siphoning downwards of the greater profits earned in the capital-intensive sector enables the bazar sector to absorb greater population.

5. Sinclair (1978) published an annotated bibliography on the topic with 126 entries. *World Development* (Vol. 6 No.9/10, 1978) published the first issue exclusively focusing on IS. UNCRD's *Regional Development Dialogue* (Vol. 2, Autumn, 1984) issue dealt entirely with IS with 248 entries in the bibliography and 13 articles. C. M. Rogers (1985) published a review on IS containing 210 entries. In 1991, the Development Centre of OECD reviewed two decades of IS definition, surveys, policies and programmes and included a bibliography of 117 entries (Lubell: 1991). In the same year, the ILO published an annotated bibliography on the IS that summarized 456 entries (ILO/PREALC: 1991) and an annotated bibliography of IS studies in Africa containing 246 annotated entries and in 1993, Asian Institute of Technology (AIT) published a bibliography containing 880 entries (AIT: 1993). Again, *Regional Development Dialogue* (Vol. 17 No.1, 1997) devoted an issue to IS. All these give some idea of the scale and quantity of work accumulated during about two and half decades of extensive research on IS (Amin, A.T. M. N: 1996).
6. Suggested criteria for identifying IS enterprises by Sethuraman (1976, Appendix, p. 81): (1) Manufacturing: A manufacturing enterprise may be included in the IS if it satisfies one or more of the following conditions: (a) It employs ten persons or less including part-time and casual workers. (b) It operates on an illegal basis, contrary to government regulations. (c) Members of the household of the head of the enterprise work on it. (d) It does not observe fixed hours/days of operation. (e) It operates in semi-permanent or temporary premise or in a shifting location. (f) It does not depend on formal financial institutions for its credit need. (h) its output is normally distributed direct to the formal consumer. (i) Almost all those working in it have fewer than six years of formal schooling.

(2) Construction: A construction enterprise may be included in the IS if it satisfies one or more of the following conditions: (a) Any of (1) (a)-(c) or (i) above. (b) It does not own power-operated construction machinery and equipment. (c) It is engaged in the construction of semi-permanent or temporary building only.
(3) Transport: An enterprise providing services related to transport, storage and communication may be included in the IS if it satisfies one or more of the following conditions: (a) Any of (1) (a)-(c), (g) or (i) above. Condition (1) (c) does not apply to transport activity *per se*. (b) It does not use any mechanical power.
(4) Trade: A trading enterprise may be included in the IS if satisfies one or more of the following conditions: (a) Any of (1) (a)—(e) above. (b) It deals with second-hand goods or sells prepared food.
(5) A service enterprise may be included in the IS if it satisfies one or more of the following conditions: Any of (1) (a) (b) above.

7. From his case study of Valsad district of Gujarat, Breman comes to the following conclusion regarding the features of the IS: (a) Government regulations concerning labour relation even the most minimal protection and security do not apply to IS. Bargaining occurs mainly on an individual basis. (b) The decisive factor in getting employment in the IS is the personal contact. The nature and the extent of this network of contacts determine not only where one is incorporated into the work process but also where, how, for how long and for what type of work. (c) Working conditions in IS have very pronounced personal overtones in the sense that there are widely varying differences between individual labourers and their employers.
8. The following is the hierarchy of informality (or formality) by Bose (1980)
   (a) Giant multinational corporation——— Formal.
   (b) National monopolies in the third world——— Both formal and informal.
   (c) Other large and medium units——— Both formal and informal.
   (d) Small-scale units——— Both formal and informal.
   (e) Tiny mainly self-employed——— Informal.
9. The important features of definition of 'corporation' and 'quasi-corporation' are: (a) Corporation: A corporation is a separate legal entity which is independent of the household(s) or household members that own it. (b) Quasi-corporation: A quasi-corporation is not a separate legal entity but it behaves like a corporation and keeps a complete set of accounts which would permit a clear distinction between the production activities of the enterprise and the other activities of the owner(s).
   Some implications of excluding corporation and quasi-corporation are: (a) Very small enterprises which are corporations or quasi-

corporations are excluded from IS. (b) The location of the enterprise is not in-and-of itself a criterion for classifying enterprises by sector. A corporation may be located in the home of the person who owns it. An unincorporated household enterprise might not be located in the home of the person who owns it. (c) The status of the employment of the persons engaged in the enterprise is not taken into consideration. The staff of a corporation may include unpaid family labourers, whilst an unincorporated enterprise may have regular paid employees with fixed wages (Govt. of Pakistan: 1994).

10. The 15th ICLS addressed several issues that could arise in determining whether particular enterprise of an employer should be classified as informal (Govt. of Pakistan: 1994). According to ICLS, in principle the size criteria should preferably refer to the number of employees employed on a continuous basis. For practical reason, the criteria of employment size may, however, also be formulated in terms of total number of persons engaged during a specific reference period (including business owners and contributing family workers). The size limit was not specified by the ICLS because it may have to vary between countries and branches of economic activity. In the case of multi-establishment enterprises, the size criteria should, in principle, refer to each of the establishments separately rather than to enterprise as a whole. The registration criteria can apply to the enterprise or its employees.

2

Rural Informal or Non-farm Sector

When we talk of informal sector, we usually mean informal sector in urban areas; and therefore termed as 'urban informal sector'. However, similar activities in rural areas are termed as rural informal sector (RIS). On the other hand, the non-agricultural activities in rural areas are defined as rural non-farm sector (RNFS) by most of the scholars. We think, the non-farm activities in rural areas may be designated as rural non-farm sector. In this chapter, an attempt is made to find out details of the features of RNFS, so that we can have a clear idea of urban informal sector in comparison to RNFS.

In the process of economic development, agriculture and the modern formal industrial sector are not able to absorb the growing labour force in the developing countries. Since the agricultural sector in the rural areas is oversaturated, the growth of RNFS is seen as a promising sector for creation of sustainable employment and as the solution to the problem of poverty and unemployment in rural areas. Until recently, the commonly held view was that RNFS was a non-productive sector which is expected to wither away with development. However, there is growth of RNFS in most of the developing countries including India. Recently much interest has been created in the nature, characteristics, structure, operational problems, policies and factors affecting the growth of RNFS.

At present, it is argued that in a situation of rural workforce growing faster than the employment potential of agriculture, RNFS has positive role to play in promoting growth and welfare. So, it is argued that that RNFS (i) can lower unemployment by absorbing growing labour force;

(ii) reduce rural-urban migration and thereby contribute in easing urban congestion and lowering pressures on scarce urban infrastructure facilities; (iii) employ women and provide seasonal employment or residual employment for those left out of agriculture and for the poor who cannot find sustenance in agriculture since they are small landholders or landless; (iv) use more appropriate technology, particularly in small-scale rural manufacturing and thus generate greater income from available productive inputs; (v) improve household security through diversification; (vi) contribute to national economic growth; (vii) provide commonly consumed goods at lower prices which are mostly demanded by the poor; and (viii) a more equitable distribution of income (Lanjouw and Lanjouw: 1995).

It is further argued that since redistribution of income through tax-transfer mechanism is politically and administratively difficult in all the developing countries, there is a case for focusing on RNFS from the point of view of income distribution. Various organisations and scholars (ILO: 1983; Liedholm and Chuta: 1979; World Bank: 1978) have written on RNFS of developing countries.

In India, after independence, the state had a declared policy of reducing or eradicating rural poverty by land reform measures. But this policy barring a few exceptions was a failure. The dissatisfaction and disgust of the rural mass was expressed in the violent form of Naxalbari movement starting from Naxalbari, Siliguri and Jalpaiguri against the feudal system and unequal distribution of land and rural assets in the late 1960s. Against the backdrop of this violent movement in rural areas against *jamindars, jotedars* and landlords, the Union Government started various rural development programmes such as FWP, IRDP and others in 1970s and 1980s to attack rural poverty, underemployment and unemployment. However, due to improper implementation and various leakages, these programmes are also not successful. Against the background of failure of land reforms and improper implementation of various rural development programmes, RNFS is seen as a sector with a potential to alleviate rural poverty and reduce unemployment.

The main focus of this chapter is a detailed and critical review of literature on RNFS, the main objectives of which are: (i) to identify the factors which are supposed to determine the level and growth of RNFS and mechanism through which these factors operate; (ii) to find out the characteristics of RNFS in developing countries including India; (iii) to examine whether it is appropriate to state the features of RNFS as a whole without taking into consideration the different sub-sectors within it; and (iv) to consider whether, in terms of spatial focus, it is appropriate to carry out analysis at the state level rather than for more disaggregated sub-divisional units.

Concept and Features of RNFS

In the background paper for the *World Development Report 1995*, RNFS is defined as the sector which includes all economic activities in rural areas except agriculture, livestock, fishing and hunting. It is not in any sense a homogeneous sector. RNFS may include: (i) activities undertaken by farm households as independent producers in their home; (ii) the subcontracting of work to farm families by urban-based firms; (iii) non-farm activities in villages and rural town enterprises; and (iv) commuting between rural residents and urban non-farm jobs (Lanjouw and Lanjouw: 1995).

Thus, RNFS includes a wide range of economic activities whose composition may vary from country to country. The most common convention is to include animal husbandry, hunting and tapping, forestry and logging, fishing, etc., in agriculture and, accordingly, all other economic activities in rural areas would constitute RNFS (Chadha: 1993). Thus, RNFS comprise all non-agricultural activities, mining and quarrying, household and non-household manufacturing, processing, repairs, construction, trade and hotelling, transport and communication, and community, personal and other services in rural areas.

Limitations of the Concept

There are certain limitations of the concept of RNFS (Lanjouw and Lanjouw: 1995). These are stated briefly below:

Different definitions of 'rural' are used in the collection of census and survey information. The definition of rural settlement on the basis of size of population varies. Thus, more limited definition of 'rural' lowers the percentage of employment in RNFS.

Various features of the data on RNFS imply that the percentage of rural employment in RNFS may be underestimated for all countries. First, the figures on RNFS in different countries refer only to primary employment. But primary employment status underestimates the actual percentage of 'labour hour' engaged in RNFS in slack season of agriculture (Haggblade et al.: 1989). Second, there is every possibility that RNFS enterprises may simply be missed by survey since they are very small and dispersed. In south African countries, under-enumeration has been as high as 40 per cent. In spite of under-estimation, RNFS is substantial in many countries, both in terms of income and employment. It has also been growing over time.

Characteristics

RNFS is not a homogeneous sector. It is a heterogeneous sector having various sub-sectors. So, it possesses various interesting characteristics. Let us look into these.

First, across the countries, the size of RNFS enterprises is very small on the whole (Haggblade et al.: 1989). They are small in terms of employment, capital investment, cost of inputs, ownership of land, premises and tools, as is found (Norcliff: 1983). In comparison to formal sector, the RNFS uses more labour intensive methods of production.

Second, RNFS comprises activities with a wide range of labour productivities. There are mainly two groups of RNFS activities: (i) low labour productivities serving as a residual source of employment, and (ii) high labour productivities (Lanjouw and Lanjouw: 1995).

There is wide variation in returns to labour in RNFS in Java (Indonesia), Uttar Pradesh (India), Thailand, and Bangladesh (Alexander et al.: 1991; Hossain: 1987; Papola: 1987; Romijn: 1987). Labour productivity in RNFS activities in China is found to have improved substantially due to infrastructure support, education, etc. (Chadha: 1992).

Third, the intensity of participation in RNFS activities is associated with population pressure. For instances, it is found from the study of 36 Nigerian villages that involvement in RNFS activities by adults and children is common and intensity of this involvement is positively correlated with seriousness of population pressure (Okafor: 1983).

Fourth, when the involuntarily unemployed do not find employment in agriculture at the prevailing wage rate, the agricultural wage is not the opportunity cost of labour. In this scenario, low wage employment outside agriculture is taken up by the involuntarily unemployed agricultural labourer, particularly those who do not have family to fall back on (Lanjouw and Lanjouw: 1995).

Fifth, in many countries, the ability of women to work outside the home is limited. Males have a comparative advantage in RNFS activities because of their ability to travel away from home and to earn a higher wage rate, while women take household responsibilities relating to cooking and child care, etc. The opportunity cost of the time of women also bears little relation to the agricultural wage. Working in cottage industry is particularly useful for women from the point of view of combining it with household work, an employment in cottage industry is most often a part-time occupation for women, as in Africa and Bangladesh (Bagachwa and Stewart: 1992; Hossain: 1987).

Sixth, whether inequality in distribution of income increases or decreases due to participation in RNFS activities is not easy to say. However, it is presumed that if a large portion of RNFS income goes to the higher income group of society, than opportunity to engage in RNFS activities increases inequality in income distribution and *vice versa* (Lanjouw and Lanjouw: 1995).

The evidence from empirical studies is mixed. In some studies, it is found that the poor and landless get a higher percentage of income from RNFS occupation in Japan, Taiwan and South Korea (White: 1992) and Kenya, Botswana, Nigeria and the Gambia (Bagachwa and Stewart: 1992). These studies imply that RNFS plays some role in bringing equality in distribution of income and alleviating poverty. On the other hand, the studies in India, Kenya and Vietnam show that the relationship between RNFS income and total income or assets is U-shaped (Evans and Ngau: 1991; Walle: 1994). Here, as the assets and income rise, the tendency of seeking RNFS work decreases thereby reducing the share of RNFS income in the total income. But after a point, as the total income and assets rise, there is a tendency to invest the surplus on RNFS activities, thereby increasing the proportion of income from RNFS activities in the total income.

Seventh, many people in rural areas are seasonally unemployed. In slack season, a rural worker may not be able to find employment in the agricultural sector. So, even a low productivity occupation with a low wage/income can be useful to raise and provide stable income over the year (Lanjouw and Lanjouw: 1995). However, the types of RNFS employment which are available on a seasonal basis are limited. Capital intensive RNFS activities are not likely to be undertaken seasonally because capital is underutilised during the agricultural peak seasons.

The seasonality of employment in RNFS activities is found in various empirical studies. The incidence of seasonality in RNFS employment is observed in Thailand (Romijn: 1987; Shand: 1986), Bangladesh (Reza: 1986) and Philippines (Fabella: 1986). But, in Nigeria, the level of involvement in RNFS activities does not seem to be subject to seasonal fluctuation (Okafor: 1983).

On the basis of seasonal variation, RNFS activities may be grouped into two categories: (i) those competing with agriculture for labour and run parallel to agriculture cycle, and (ii) those that are complementary to agriculture and are taken up during agricultural slack period and linked to agriculture through market demand (Fabella: 1986).

Eighth, most of the households earn income from multiple sources and have multiple jobs. RNFS income may stabilise income by spreading risk through diversification. When agricultural output varies greatly over the year and across the year and where mechanism for stabilising income such as credit and transfer is absent, RNFS plays an important role in risk diversification.

Inter-Linkage Hypothesis and India

The linkages between agriculture and the rest of the economy have drawn the attention of the economists from the time of Physiocrats. Hirschman (1958) advocated unbalanced growth in his theory of 'big push' with particular reference to industries. While Hirschman analysed two types of linkages between industries, Watanabe (1983) classified linkages between industries into five categories. But to Hirschman, compared to industry, agriculture is a weaker stimulant to the starting of new economic activities through linkage effects.

However, some studies (e.g., Mellor: 1976) show that agriculture has the capacity to stimulate new economic activities in rural non-farm sector (RNFS) through consumption-expenditure linkages. Various other types of growth linkages such as: (i) backward linkages, (ii) forward linkages, (iii) capital flow linkages, and (iv) labour flow linkages of agriculture are also recognised (Haggblade et al.: 1989; Harriss: 1987).

In the mid-1970s, John Mellor (1976) stated the role of RNFS in a set of proposals for India. The growth of farm and non-farm sector would be mutually reinforcing with employment and income increasing in a dispersed pattern. In his view, geographic isolation and tastes of rural population combined to make demand for locally produced goods increase with income. It is argued by Mellor that due to growth in productivity in agriculture, income from agricultural sector rises which consequently expands the consumption demand of farmers for labour-intensive goods and hence stimulates the growth of RNFS.

According to the argument of Mellor (1976), technological innovation which brings increased food supply normally distributes the initial benefits largely to already more wealthy and prosperous class. This initial increase in rural income sets in motion a sequence of multiplier effects which can stimulate expanded production and employment in other sectors of the economy including consumer goods industries and small-scale units in RNFS which are likely to be labour-intensive. The expanded income due to higher employment of lower-income labouring class, who spend larger portion of their increased income on food, provides the demand for additional foodgrains production. In this way, the inter-sectoral linkages between farm and non-farm sectors in rural areas are established which are beneficial to each other and stimulate both sectors.

So, the focal theme of Mellor's argument (the inter-sectoral linkage hypothesis) is that without having any equal distribution of rural income and assets, there can be growth through inter-sectoral linkages between farm and non-farm sectors in rural areas which can only be achieved by development of capitalist agriculture through introduction of technological advance.

Agricultural Development as a Stimulant

Some studies in India suggest that the growth in agricultural production is likely to stimulate the growth and development of RNFS (Bhalla: 1990, 1993; Dev: 1990; Papola: 1987, 1992; Rao: 2006; Shukla: 1992; Unni: 1991, 1994). In his study on India, Papola finds that in different states, the performance of rural industrial sector, a sub-sector of RNFS is associated with agricultural productivity and has higher correlation with the growth rate of agricultural output. Agricultural growth has influenced RNFS directly by supplying raw materials and creating demand for inputs and allied services and indirectly through raising consumption demand and generating surplus for investment (Papola: 1987, 1992).

Papola (1987) also argues that the major part of the rural industrial activity in different states has continued mainly as a part of the tradition without being necessarily differentiated on the basis of linkages and integration with the local resources and changing demand pattern. Due to this reason, most of the industrial enterprises in RNFS in India are carried out as a means of family subsistence rather than as business (Ibid.).

There was a significant growth in RNFS activities in the agriculturally developed state of Haryana during the three decades from 1970 (Sardana et al.: 1995). It was observed that the share of RNFS employment was the highest in the agriculturally developed districts. But simultaneously, RNFS employment has also increased significantly in a majority of districts of Haryana which are not agriculturally developed (Ibid.).

On the other hand, the assumption that the growth of RNFS is led by agricultural development is not observed in various empirical studies in India. The linkages and relationship between farm and non-farm sectors in rural areas are mixed and complex. It is observed that the level and growth of land productivity, per capita agricultural output, non-food crops, etc., are not significantly related to RNFS in taluk and district levels in Gujarat (Basant: 1994). Similarly, in West Bengal, it is found that in the districts where agricultural output grew at a rate greater than the state average, the proportion of male RNFS workers has declined during the period 1961-81 (Chandrasekhar: 1993). This study does not find any significant relationship between the shares in total rural employment in RNFS activities and the rate of growth of agricultural output and its components in 1981. The agriculturally more developed western Uttar Pradesh has experienced a shift away from RNFS due to continued ability to absorb more labour in agriculture (Singh: 1994).

Thus, it is difficult to say whether RNFS expands or contracts due to rise in agricultural productivity.

Other Factors along with Agricultural Growth

It is sometimes observed that the growth of RNFS is due to various factors including agricultural growth (Eapen: 1994, 1995; Shukla V.: 1989, 1992; Singh: 1994; Unni: 1991). In Kerala, a high degree of commercialisation of agriculture, close rural-urban linkages, rapidly declining land-man ratio, increase in the proportion of marginal holding, overall rural prosperity, flow of remittance into rural areas, and growing level of literacy have historically played an important role in generating a high level of employment in RNFS (Eapen: 1994, 1995).

In Uttar Pradesh as a whole, the level of agricultural prosperity, inequalities in income/expenditure and land distribution, mechanisation of irrigation and other agricultural operations are found to exercise a positive influence on the growth of RNFS (Singh: 1994). A large size of landless labourers and limited possibility of further absorption of labour in agriculture are associated with a higher level of RNFS employment in Uttar Pradesh. However, it is not necessarily a distress employment phenomenon since poverty levels show a strong negative association with RNFS employment (Ibid.).

Thus, the stimulants to the growth of RNFS can come from both agriculture or outside it. However, agricultural development may improve the efficiency of RNFS enterprises without necessarily increasing their number (Papola: 1992; Papola and Mishra: 1980).

Traditional RNFS

While discussing or analysing RNFS and its linkages with farm sector, the dichotomy in it should be recognised. RNFS can be classified as follows (Samal: 1997a):

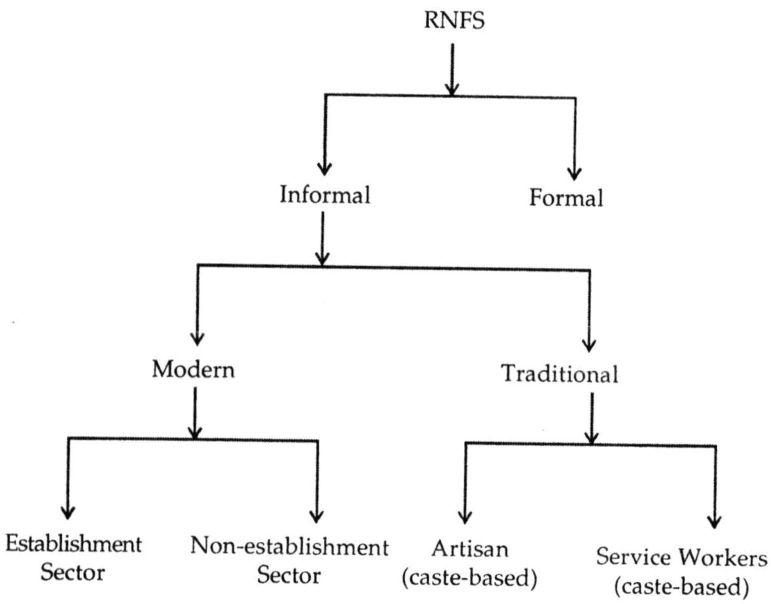

If we classify in this manner, the growth pattern will be different for different sectors. It is more likely that agricultural development may have an adverse effect on traditional RNFS, particularly on artisans. Traditional service workers (e.g., barbers, washermen, etc.) may not be adversely affected, though *jajmani* system is more likely to be replaced by market mechanism (Ibid.).

Of course, one theoretical study (Mukhopadhyay: 1985) has recognised the bio-modal nature of RNFS, such as: (i) Rural Non-Agricultural (RNA) sub-sector-I, and, (ii) RNA sub-sector-II. The first group comprises products and/or activities which are run on more or less stable basis with an eye on surplus generation and growth, using primarily hired labour and a certain degree of technological sophistication. The second group comprises products and/or services which often, though not always, seasonal are run with the help, primarily of unpaid family labourers, using rather primitive technology, catering mostly to the local market and responding more to the supply side of the labour market than the market demand for the product. Kumar (1993) has also recognised that both traditional and non-traditional components of RNFS may have different kinds of inter-linkages with agriculture.

It is to be pointed out that traditional agriculture uses mostly locally produced inputs. The large part of these locally produced inputs is supplied by agriculture and animal husbandry. The dependence of agriculture using modern technology on non-local inputs is comparatively more. Thus, while growth of agricultural production generally accounts in a more than proportionate increase in demand for non-agricultural inputs, it does not necessarily lead to an increase in demand for locally made inputs as a whole (Vaidyanathan: 1994). Perhaps, for this reason, the number of workers employed in traditional RNFS in India has declined or remained stagnant while there is a corresponding expansion in the non-traditional sector.

It is found that though small-scale traditional manufacturing mostly in rural households has been protected and subsidised in India, it has declined relatively (Little: 1987). Similarly, traditional industries in RNFS diminished rapidly in importance in other developing countries such as Malaysia, Philippines, Taiwan, Pakistan, Colombia, etc. (Lanjouw and Lanjouw: 1995; Little: 1987). Of course, in some rural areas in India, traditional RNFS still continues to cater to local consumption needs and to small production requirement of agriculture (Papola: 1987).

The same declining trend of traditional RNFS as well as presence of disguised unemployment and declining output per worker in it are also observed in empirical studies on various states like Gujarat, Himachal Pradesh, Haryana, Uttar Pradesh and the eastern region of India (Basant: 1994; Sardana et al.: 1995; Sarswat et al.: 1995; Singh: 1994; Verma and Verma: 1995).

In our study (Samal: 1997a, 1997d) of village Pahadasigida, one of the eight villages having concentration of artisans of the most prosperous agriculturally developed block of Orissa, viz., Attabira, it is observed that there is a fall in the sale of products of traditional RNFS consisting of three categories of artisans, viz., blacksmith, carpenter and potter, which is either due to the availability of substitutes produced by modern formal industry, mostly located in urban areas (40 per cent of cases) or due to availability of similar products of big formal industry at cheaper price (30 per cent cases). Moreover, in the past, a sizeable portion of the demand for the product of artisans (traditional artisans) in the block was coming from large farmers but it has come down sharply (from 60 per cent to 37 per cent) at present. The main portion of demand (46 per cent) for artisans' products at present is from small farmers. But none of the respondent artisans has changed his occupation over the last 30 years.

Thus, though there may be a rise in the number of people engaged in traditional informal RNFS, the sale and income of this sector have declined over time validating the inference that there is rise in disguisedly unemployed in traditional RNFS. But it is not correct to presume that the disguidely unemployed of the agricultural sector have moved out of it and joined as disguisedly unemployed in traditional RNFS (Ibid.).

On the other hand, in the industrial belt of Talcher-Angul of Orissa, we found that income from agriculture and allied sector as a percentage of total income has declined from 45 per cent in 1983 to 20 per cent only in 1994; while the importance of RNFS as a source of household income has been growing, the proportion of this income increasing from 55 per cent to 71 per cent during the same period (Ibid.). However, most of these incomes are from modern formal sector. This is because of the rise in number of service holders in public sector undertakings and government sector, particularly in NALCO, since the affected persons of these villages are employed in NALCO under rehabilitation policy and some others in other non-farm activities coming out due to linkage effects.

Inequitable Distribution of Land: A Constraint

We have observed that the growth of agricultural development has led to decline in traditional RNFS, particularly in manufacturing, in most of the rural areas of India. That does not necessarily imply that it led to the growth of modern RNFS everywhere. However, the growth of labour-intensive RNFS is seen as the solution to the problem of rural unemployment and poverty. So, it is argued that inter-sectoral linkages between agricultural sector and RNFS would generate the growth of RNFS by generating productive employment (e.g., Mellor: 1976).

But this argument does not realise that the inequitable structure of rural economy and inequitable distribution of land and rural assets are constraints to the growth of RNFS. It is empirically found that in the green revolution belt of Punjab, large farmers are not the major source of demand for labour-intensive RNFS goods and services (Ray: 1994). Technological development in agriculture led to rise in production expenditure on non-local inputs, rather than consumption expenditure. In general, rising income due to growth in agriculture generates demand for non-local goods.

Even modern technology in agriculture and commercialisation of agriculture accentuate the inequality in income distribution of rural society by adversely affecting the sub-group rural poor and this inequality, in turn, hardly stimulates RNFS as a whole through consumption-expenditure linkages. A general equilibrium model shows that income gains from green revolution in India initially accrued to the wealthier rural groups, but that after 1972-73, they were transferred to consumers mostly in urban areas and, that by 1980-81 real per capita income of rural population was barely above their 1960-61 level (Quizon and Binswanger: 1986). In the early green revolution period, large farmers gained the most while the rural poor gained very little. But the sub-group of rural poor including the landless agricultural workers suffered severely.

Thus, the gains in agricultural production during the late 1970s were not reflected in further advances in rural income because the prices of agricultural products fell. Instead, the productivity gains were transferred to consumers mostly in urban areas by way of declining prices. Thus, the increased production expenditure of the rich farmers and gains in productivity in terms of lower food prices accrue to the urban centres. In this sense, the rise in agricultural productivity may have little impact on RNFS.

In reality, technological innovation and commercialisation of agriculture have accentuated the inequality in the rural society which, in turn, may not facilitate the growth of RNFS. In the real world experiences, the poor have failed to reap the benefits or even lost from the technological change or commercialisation. Where these benefits have occurred, it is found that they are mostly attributable to inelastic demand or adverse institutional features (Binswanger and Braun: 1991).

Even the income of the wealthier farmers who gained from the rise in agricultural production may not help in growth of RNFS. The consumer products such as cosmetics and toiletries, etc., and, agricultural inputs like fertilisers, pesticides, implements, etc., are produced by large industrial firms including multinationals in the formal sector mostly situated in urban centres. Even if these goods are produced by non-traditional RNFS, they may not find market among the rural wealthier class who have a psychological preference for products produced in big cities by formal sector units.

Thus, the rise in consumption demand for these products emanating from the increased income arising out of the agricultural sector generates demand for the products of mostly urban-based, capital-intensive formal manufacturing sector including multinationals. Hence, the surplus generated in agricultural sector, instead of being spent on RNFS goods, is siphoned off to large industries in urban centres and, in some cases, to foreign countries through large dividends and frequent issue of bonus shares by multinationals (Samal: 1997a).

As mentioned earlier, similar is the case with traditional RNFS. Due to rise in agricultural income, the tastes of the people change and they prefer the substitutes for these RNFS products produced in urban centres such as cement, iron rods, corrugated iron sheets, plastic pots, aluminium and stainless steel utensils, factory-produced synthetic and cotton clothes, cycles, wristwatches, radio, bicycle, etc. in place of *khaparali* and tiles, pottery and bell-metal utensils, handlooms, bullock carts, etc. Thus, the competition from the formal sector mostly situated in urban areas may lead to decline of traditional RNFS whose products may not get a ready market with large capitalist farmers (Ibid.).

It is pertinent to mention that in Taiwan, agriculture gave the initial impetus to the development of rural industry and this development was stimulated by a fairly equitable distribution of rural income, land and assets and investment which, in turn, contributed to an annual growth of 11.5 per cent in RNFS income over the period 1962-80 (Lanjouw and Lanjouw: 1995; Ranis and Stewart: 1993). Perhaps, due to comparatively more equal distribution of land by vigorous implementation of land reform measures, the percentage of RNFS workers among Rural Main Workers (RMW) was highest in Kerala followed by West Bengal in 1980s. In one study relating to rural Karnataka, it is found that the distribution of agricultural land and assets are positively associated with participation in RNFS activities (Rayappa: 1986).

Liberalisation and Neo-Colonialisation

Mellor's (1976) assumption that both geographic isolation and the unchanged tastes of the rural people will stimulate the demand for locally produced goods may not be fully correct in this age of privatisation and liberalisation leading to globalisation. Under liberalisation and gobalisation, some of RNFS activities in developing countries may decline. The entry of multinationals and the liberalisation of imports in India, might lead to neo-colonialistic tendency. In this sense, the model formulated by Hymer and Resnick in the 1960s and recent extension of the model by Ranis and Stewart to explain the purported decline of RNFS activities under colonialism are very much appropriate for India (Hymer and Resnick: 1969; Ranis and Stewart: 1993).

With the advent of either direct colonial links or neo-colonialism through entry of multinationals and liberalisation of imports and exports, two trends are observed: (i) emergence of new opportunities for exporting cash crops and natural resources and, (ii) availability of cheap and higher quality manufactured goods from foreign countries. Both, the competition from imports and drawing off of labour into the growing cash crop sector, would stifle RNFS activities.

When heterogeneity of RNFS is recognised, some parts of RNFS will be dynamic while others will be declining (Ranis and Stewart: 1993). Colonialism stifled RNFS in Philippines. The imported goods under colonialism in Java (Indonesia) had a detrimental impact on some sub-sectors of RNFS (Ibid.).

Similar is the case in India. Traditional RNFS and the rural economy in India are, at present, being shaped by uncertainty and transience which are further fuelled by the process of liberalisation and increased reliance on market forces. In the past, during the British colonial period, there was disruption of traditional RNFS activities brought about by a combination of merchant power and state power (Balkrishna and Gopalkumar: 1995). There was also simultaneous transformation of social ethos and shift in tastes and preferences of the consumers.

Similarly, at present, as a result of liberalisation, and transition to market-determined economy and the consequent entry of multinationals and transnationals in India, there will certainly be a very dampening impact on traditional RNFS due to flooding of cheap substitutes and similar products and change in tastes and preferences of both the rural and urban consumers on account of aggressive advertisement in both print and multi-channel electronic media of their products (Samal: 1997a). Thus, the rise in agricultural productivity, if any, may not expand RNFS activities under the regime of liberalisation and implementation of WTO regulations.

Other Hypotheses on Determinants of RNFS

Another view is that RNFS is a residual sector since the workers in the rural areas who are not able to find employment in the agricultural sector are to join RNFS. It is also argued that factors responsible for the growth of RNFS do not lie wholly in agriculture or in rural areas.

Residual Sector Hypothesis

It is argued that the growth of RNFS is caused by overall development. It is stated that the rural workers who are not able to find employment in the agricultural sector are forced to join RNFS. So, RNFS is regarded as the residual sector.

However, some empirical studies in India strongly belie the residual sector hypothesis while others support it. Bhalla (1990) has identified two kinds of distress situation in which RNFS becomes residual labour force. The first is supplementary workers who have no main occupation but engage in subsidiary work to supplement household income. The second refers to persons with a main occupation who also engage in secondary activity. RNFS activities in case of eastern region of India and eastern UP appear to be residual activities in rural areas (Singh: 1994; Verma and Verma: 1995). It is argued that excess labour pushed out of agriculture has no alternative but turn to RNFS in search of job. But the assumption that the disguisedly unemployed of agriculture join traditional RNFS as disguisedly unemployed, which is mostly caste-related skill-based, and are unable to join the urban informal sector, may not be a reality.

On the other hand, the rise in rural real wage rates across the board taken together with the rapid expansion of RNFS employment in India does not validate the residual sector hypothesis (Vaidyanathan: 1986, 1994). Similarly, at the state level also, the growth of RNFS employment is not necessarily due to distress condition (Basant: 1994; Dev: 1990; Unni: 1991).

The most important effects of distress condition in the rural areas is neglected by the residual sector hypothesis. Due to unfavourable conditions in rural areas, such as prevalence of capitalist production in agriculture, decline of rural handicrafts, inadequate income, poverty, unemployment, underemployment, seasonal employment, loss of property and sources of income due to natural calamities, etc.; the landless agricultural labourers, small marginal farmers and artisans are pushed out of the rural areas and move mainly to urban centres in search of jobs in informal sector (Samal: 1997a).

In contrast to the Lewis-Ranis-Fei model of one-way migration (Lewis: 1954; Fei and Ranis: 1967), the rural-urban migration is regarded as two-stage phenomenon in Todaro model. Todaro (1969) recognised a dichotomy in the urban economy. In the first stage, the unskilled rural worker migrates to an urban area and initially spends a certain period of time in the so-called urban traditional sector. In the second stage, after a certain period of job search, he eventually attains a job in more permanent modern sector.

While Todaro gives importance to 'pull' factors in the process of rural-urban migration; others (e.g., Amin, S.: 1974) emphasises 'push' factor—a force of expulsion from the rural areas. Thus, due to distress condition or 'push effect', the workers who are pushed out of the agricultural sector may not be able to start any RNFS enterprise due to lack of capital or may not find year-round employment in RNFS. They may move to join the urban informal sector for year-round employment, higher rural-urban current earning differential and other social and economic disadvantages (Papola: 1981; Samal: 1990).

Of course, a study of semi-arid villages (of Gujarat) has taken the aspect of migration into consideration by bringing out distinction between categories of RNFs activities (Shylendra and Thomas: 1995). One is the non-farm proper (NFP) which includes: (i) artisans/services, (ii) trade/business, and (iii) white collar and other jobs, movement towards which is largely due to increased local demand for different goods and services or the expansion of government activities The second is that of non-farm migratory (NFM) activities. Most of the households in the village have been compelled to migrate during lean-periods as a result of distress.

Thus, though the study has taken the seasonal migration into account, it has not dealt with rural-to-urban migration in general and joining the urban informal sector by rural migrants for their year-round work and livelihood which is what is observed in other areas.

Urbanisation Hypothesis

The process of urbanisation also influences the growth of RNFS and sometimes excercises a positive impact on RNFS employment (Basant: 1994; Jayaraj: 1994; Kundu: 1991; Shukla: 1992; Singh: 1994; Unni: 1991; Visaria and Basant: 1994). The role of locational factors is also emphasised in the determination of rural non-farm employment possibilities. While rural non-farm jobs are predominantly located close to urban areas, rural non-farm self-employment jobs are geographically dispersed around the country, depending on local factors, such as a profitable agricultural activity, an important road, or a tourist attraction (Isgut: 2004).

Other Factors

Besides the residual sector hypothesis and urbanisation hypothesis, there are also other arguments which mention various factors that influence RNFS activities (as observed by the various empirical studies in India) such as:

(i) changes in the tastes of the rural consumers,
(ii) competition from factory sector (Visaria and Basant: 1994),
(iii) administrative, development and social services,

(iv) rural electrification (Singh: 1994),
(v) flow of remittance to rural areas,
(vi) growth in consumer demand (Eapen: 1994),
(vii) literacy rate of the population (Basant:1993; Jayaraj: 1994; Rayappa: 1986),
(viii) proximity of rural areas to urban areas (Basant: 1993; Jayaraj: 1994; Rao: 2006),
(ix) level of development of transport system (Jayaraj: 1994; Rao: 2006),
(x) the level of rural and extra-local demand (Vaidyanathan: 1994),
(xi) caste (Rayappa: 1986),
(xii) landlessness (Basant: 1993),
(xiii) industrial development (Mehta: 2003, Samal: 1997d).

Features of RNFS in India: A Review

The size of RNFS in India is measured in terms of employment using Census of India and NSS data. RNFS includes all economic activities other than crop production and allied agricultural activities such as animal husbandry, plantation, fishing, etc. It includes both principal and marginal workers. The following features of RNFS in India have been observed by various studies.

First, RNFS workers as a percentage of rural workforce in India has been increasing. RNFS incomes have also been rising steadily (Chadha: 1993). It is observed that some of RNFS activities rise with the size class of the village (Visaria: 1995).

Second, within RNFS in India, the increase in the tertiary sector seems to be higher than that in the secondary sector (Basant and Kumar: 1994). The growth in employment in manufacturing sector of RNFS is modest because of continuing decline in the household industry and the substitution of capital for labour (Visaria: 1995). In the rural industrial sector of RNFS, household production is more prominent in most consumer goods industries (Mathur and Pani: 1993). Attainment of higher level of rural development tends to promote expansion of non-household component of rural industrialisation but a relative shrinkage of the traditional household industries. The degree of spatial spread is more in non-household segment than within the household segment for most rural industries. The growth of employment in the construction sub-sector of RNFS in India was partly a result of scarcity relief work initiated in states (Visaria: 1995).

Third, most of the increase in RNFS is explained by the increase in the proportion of casual RNFS workers and, the casual agricultural labourers report a much higher incidence of RNFS work in a secondary or subsidiary capacity (Basant and Kumar: 1989). Most of the seasonal fluctuation in RNFS employment can be explained by the change in employment structure of rural casual labourers who shift back and forth between farm and non-farm sectors.

Fourth, the sources of RNFS employment are far too scattered in space (Chadha: 1993). This type of employment is available to workers in their own village, nearby villages or in the nearby town.

Fifth, in rural areas, there are no water-tight compartments of employment activities. Some households whose declared principal occupation is agriculture do engage in other activities. Cross-engagements are common. The important features of this cross-engagements are: (i) The household whose declared principal occupation is in any sub-sector of RNFS does also report involvement in agriculture. (ii) The cross-engagement is largely in the form of wage-paid employment (Ibid.).

Sixth, the extent of participation in RNFS activities of the members of the household is determined by the size of the holding, besides the ownership of land (Rayappa: 1986). As the size of the holding goes up, the tendency of seeking RNFS work decreases. The landless and marginal landholding households spend a higher proportion of their employment time in RNFS jobs (Chadha: 1993). It is observed by a study that as the size of the landholding becomes smaller, the proportion of RNFS households increases consistently. That is, an inverse relationship between farm size and the proportion of rural households engaged in RNFS is observed in India.

Seventh, there has been now considerable rise in the proportion of women to men workers in manufacturing activities of RNFS in India. This is not necessarily due to replacement of male workers by female workers leading to rise in male unemployment nor a reflection of substitution of low-paid female workers (Mitra: 1993).

Eighth, education is regarded as a crucial input for employment in RNFS in India. Proportion of illiterate workers in RNFS is much less than that in agriculture (Chadha: 1993). Educational attainment facilitates a shift from agriculture to RNFS since educational qualification is helpful to get RNFS jobs.

RNFS in Orissa: Some Tentative Features

A study group appointed by NABARD has observed certain features of RNFS in Orissa (NABARD: 1994). According to their definition, RNFS comprises all non-agricultural activities: mining and quarrying, household and non-household manufacturing, processing, repairs, construction, trade, transport and other services in villages and rural towns upto 20,000 population, undertaken by enterprises varying in size from household 'own-account enterprise', all the way to 'factories'. On the basis of this definition, the following features of RNFS of Orissa are observed (Samal: 1997a).

First, there are regional variations in the importance of RNFS to the local economy within Orissa. There is also an element of sub-sector-specific clustering in RNFS. Around two-thirds of paddy processing units are located in undivided Sambalpur district only, cotton textile in Cuttack, Sambalpur and Ganjam area, spice processing in Ganjam district, fibre products and stoneware in coastal districts, non-timber forest products collection in Koraput and Kalahandi and mineral-based industries in Dhenkanal, Keonjhar and parts of Sundargarh district. Thus, the specific clustering is mainly due to the availability of raw materials and inputs. However, cotton textile concentration in Ganjam, Cuttack and Sambalpur areas is on account of long strong handloom tradition in the region (Ibid.).

Second, major portion of RNFS workers are engaged in household manufacturing and other services. There has been increase of rural main workers in household manufacturing sub-sector of RNFS in Orissa. But, manufacturing in the non-household sector increased slowly. Some of the high growth RNFS sub-sectors of Orissa are: (i) services, (ii) electricity and water and, (iii) jute/hemp products (Ibid.).

Third, a large number of artisans and cottage level producers in traditional RNFS use traditional technology while a small number of large units like rice mills use modern technology.

Fourth, availability of credit for both fixed capital and assets and for working capital is the major problem for most of the sub-sectors of RNFS in Orissa. A large majority of RNFS producers had to rely on informal credit sector due to inadequacy of loan amount, unnecessary delay and excessive paper work in getting a loan from formal sector financial institution (NABARD: 1994).

Fifth, the sustained growth in both domestic and export demand is found to be an important factor determining the growth of a sub-sector of RNFS in Orissa. The domestic demand seems to have helped the growth of sub-sectors such as spice processing, cotton textiles, minor mineral processing, handicrafts, repair and rural transport. On the other hand, the growth of textile products, cut flowers, handicrafts and handloom products is influenced by the export demand. But, there is constraint in the growth of sub-sectors like pottery and wood products due to lack of demands.

Sixth, there is dominance of artisan-based units like filigree, bell-metal, stoneware, sawai-rope making in the traditional industrial profile. In traditional RNFS, the government policy in protecting the employment of handloom weavers is quite prominent. Most of the handloom weavers are part-time workers and do not get gainful employment round the year. The demand for handloom has been declining. Most of the traditional RNFS industries are facing the same challenge (Samal: 1997a).

Determinants of RNFS in Orissa: A Static View

It is observed that though the percentage of rural families below poverty line was highest in Mayurbhanj district, the district ranks ninth in terms of percentage of rural main workers of the district engaged in RNFS. Hence, the presumption of residual sector hypothesis that the poverty of the people due to agricultural distress forces them to join RNFS may be open to question (Samal: 1997a).

Land Distribution

The rank correlation coefficient between the percentage share of RNFS workers to rural main workers (RMW) and equal distribution of land is positive and significant in Orissa (Ibid.). This implies that more is the equal distribution of land, larger is the percentage share of RNFS workers in RMW of the district.

This may be due to the fact that a more equitable distribution of land and rural assets, say, under an effective land reform policy, imply the transfer of rural assets and thereby the income and purchasing power from the wealthier peasants to landless and marginal farmers. This, in turn, may transfer the rural consumption demand from non-local products, produced by large industrial houses including multinationals mostly in the urban formal sector, to local products produced in RNFS, considering that the wealthier, capitalist farmers gaining from agricultural productivity due to improved technology and commercialisation of agriculture are not the major source of demand for RNFS products.

Thus equal distribution of land and rural assets including common property resources and water for irrigation and RNFS activities are positively related through consumption-expenditure linkages (Ibid.). However, whether inequality in distribution of income increases or decreases due to participation in RNFS activities, is not easy to say.

Poverty, Agricultural Productivity and Urbanisation

It is found that coefficient of correlation between the percentage share of RNFS workers (RNFSW) in RMW and the percentage of rural families below poverty line are negative and significant. It implies that districts having lower percentage of families below poverty line have higher percentage share of RNFSW in RMW (Ibid.). Thus, RNFS in Orissa is not a residual sector. On the other hand, RNFS activities might have helped in alleviating poverty in the rural sector.

It is also observed that the agricultural productivity has little to do with RNFS activities. Therefore, it is not possible to predict *a priori* whether RNFS employment expands or contracts with agricultural productivity. On the other hand, the percentage of urban population in the district positively affects the share of RNFSW in RMW (Ibid.).

Literacy and Education

Most important observation from the study (Ibid.) is that the districts with high literacy rate have comparatively larger percentage share of RNFS workers in rural main workers. Except some extreme cases, the districts with high percentage of urban population have high literacy rate. This may be due to comparatively better availability of educational facilities in urban centres, the benefit of which accrues to both urban and nearby rural areas. Thus, urbanisation may positively influence the literacy rate and educational attainment, which in turn may positively influence the share of RNFS workers in rural main workers. On the other hand, the districts with higher percentage of rural families below poverty line have lower literacy rate.

Thus, literacy rate is positively influenced by the level of urbanisation and negatively by poverty in Orissa. However, RNFS is strongly influenced by literacy rate and educational attainment. The positive relationship between RNFS and literacy is also observed in other states (Basant: 1993; Jayaraj: 1994; Rayappa: 1986).

Education, Income Distribution and Employment

Education and literacy might help in growth of RNFS since it improves the skill of a person which is required to start a modern RNFS enterprise with new technology or high labour productivity activities. Moreover, any education, at least beyond seventh standard may discourage a person to work as agricultural worker or owner-operator cultivator, since agriculture is regarded as dull, monotonous and low-status job by the educated in Orissa—to plough as bullock driver or transplant in muddy water. So, an educated person after completion of his education joins RNFS, rather than working in agriculture, after his waiting period, during which he searches for a better job in RNFS, either as self-employed or wage worker or in a traditional activity under market mechanism (Samal: 1997a).

Education has a large and significant impact on productivity growth. That is, literacy and educational attainment help in skill formation which, in turn, expands the growth of RNFS. Moreover, educated and literate men have more access to information and available facilities including inputs and assistance from public institutional systems which helps to start and expand a RNFS enterprise. The RNFS enterprise requiring modern technology is easily adopted by the educated and literate.

It is also observed from the study of 25 of the world's largest countries that spread of technology depended on the learning potential and motivation that are linked to the development of formal schooling (Psacharopoulos: 1988). It is to be pointed out that no nation with uneducated and illiterate citizens is developed. However, the rate of return is the highest in primary education followed by secondary and then university level. Educational attainment and literacy also affect positively efficiency in resource allocation leading to higher income and more equitable distribution of such income (Ibid.).

Since education has strong impact on individual's earnings, the net effect of the expansion of schooling has been a reduction in the dispersion of earnings and hence a more even distribution of income. However, this equity effect depends on the level of schooling expansion. The basic and primary education has the highest impact on distribution of income favourable to equity, while equity impact of expansion of post-graduation may actually be negative (Ibid.). This is because low level of earnings of otherwise illiterate workers are raised to the overall mean by imparting to them primary education, while post-graduate education received mostly by high-income groups in society will raise their incomes further away from the mean.

Thus, reduction in both, the dispersion of landholding even by direct measures like land reform and in the dispersion of earnings through spread of primary education will bring about less unequal distribution of income and may help to boost rural income and assets and thereby RNFS activities through consumption-expenditure linkages.

Thus, education (primary and secondary) not only helps directly in the growth of RNFS activities by diverting workers from farm to non-farm sector and helping skill formation but also helps indirectly through consumption-expenditure linkages by bringing about less unequal distribution of income in rural society (Samal: 1997a).

Conclusion

The three past well-known hypotheses argue that the level of RNFS is either determined by agricultural development, or urbanisaton or forced expulsion of labourer out of agriculture. But is it correct to presume that coming up of a large-scale non-agro-based industry or mining activities or public services in rural area is determined by these factors?

It is imperative to think of dualism in RNFS. RNFS can be classified as formal and informal, and a further distinction can be made between traditional and modern within informal RNFS. If at all, the factors mentioned in the hypotheses determine the level of RNFS, these may influence the informal sub-sector rather than formal component of RNFS, except of course agro-based industrial units.

Moreover, the analysis of RNFS with reference to specific regions may give more insights than talking in terms of rural areas of a state as a whole. If an area though not in the vicinity of a town is situated in an industrial belt, RNFS is more likely to be vibrant. On the other hand, the proportion of RNFS workers to the main workers in an agriculturally and industrially underdeveloped region may be higher due to some historical reasons and strong tradition. The level and pattern of growth of various sub-sectors of RNFS are also expected to be different in different regions.

Sometimes, the level and location of RNFS, particularly service sub-sector including public administration, education and modern health services, may be determined by exogenous factors such as state policy on administration and social sector.

It is difficult to estimate the exact number of workers in RNFS due to multi-occupational status, different definitions of 'rural' and under-enumeration of RNFS enterprises. The extent of secondary employment in RNFS is quite extensive. So, it is more scientific to calculate labour-hours spent in RNFS or percentage of RNFS income in total income of a family.

In spite of these limitations and difficulties in the methodology, it is observed from our field study (Samal: 1997d) that income from RNFS as a percentage of total household income has increased rapidly in the industrially developed region of Orissa. On the other hand, agricultural development in a region has adversely affected the sale and income of traditional informal RNFS, though the number of persons engaged in this sub-sector has increased implying the incidence of disguised unemployment in it.

In our analysis of RNFS as a whole in Orissa (Samal: 1997a), we come to conclusion that more even distribution of land and rural assets including water from irrigation and equal access to common property resources as distinct from subsidisation of mechanisation of agriculture and, expansion of primary and secondary education rather than subsidisation of higher education, may help the growth and expansion of RNFS activities by increasing rural income and improving distribution of such income. Moreover, education and literacy help in increasing the productivity and skill of workers which, in turn, mostly stimulate the growth of modern informal RNFS with new technology.

Thus, more even distribution of land and rural assets, as also literacy and education are the two most important factors determining the level and growth of RNFS activities, particularly the informal component of it.

Thus, to alleviate rural poverty and unemployment, the growth of RNFS is desirable which is possible by reducing both dispersion of land and rural assets by land reforms and dispersion in rural income by expansion of primary and secondary education.

3

Linkages

Mainly, linkages between informal and formal sectors have been interpreted in two ways. One view is that IS would be benefited through strengthening of linkages between two sectors. In contrast, others view that FS is an obstacle to the growth of IS because the latter is exploited by variety of mechanism.

It is observed that informal sector subsidises part of the cost of formal sector in developing countries, enabling the latter to pay comparatively low wages to their labourers. The basic needs of the workers in FS are partially met by relatively cheap goods and services produced by IS employing low-paid workers and unpaid family labourers. This subsidy consumption of FS workers is the most important role played by informal sector (Portes: 1978). FS not only benefits from IS for the maintenance of the low wage rate but by supplying casual labour—a product of highly competitive nature of IS production and high failure rate (Gerry: 1978) and some of its output which may be intermediate product for FS units (Richardson: 1984). Thus IS plays the role of a producer of wage goods and services for FS firms and of a 'reserve army' setting the supply price of labour to FS (Godfrey: 1979). This is the reason for which the elimination of IS is not sought by FS. In view of this, the optimum strategy of FS is to keep IS at an optimum size; not too big so as not to represent dangerous competition to FS but not so small as to have no influence on wages (Davies: 1979).

On the other hand, IS also depends on FS for its inputs and raw materials. Though IS plays a complementary role, it is at disadvantage when in competition with FS in various ways, viz., unit cost, advertisement, product differentiation and protection (Godfrey: 1979). Even when informal sector is complementary to FS, the former is exploited by the latter. The two sectors operate in basically two different markets. There is exploitation of IS by FS in manufacturing activities with skilled and sophisticated products mainly by controlling the marketing system (Bose: 1978; Romatet: 1983).

At present, it is observed that linkage between informal and formal manufacturing sectors is encouraged but its exploitative nature is not checked. In the case of IS units with linkages with formal manufacturing sector, cautious and careful policy of utilising the growth potential as well as regulating the arrangements so as to minimise the exploitative features of linkages need to be worked out. It is rare that linkages between the two sectors, as in the case of sub-contracting help in transfer of technology or in economic development in general, as commonly believed. So, it is essential to ascertain the relationship between the two sectors.

Theories of Linkages

The sector, which is now designated as informal, murky, grey, marginal sector and so on, has been in existence in the past; the plight of migrant workers—surplus people in 'latent' form—engaged in cart pulling, begging, manure collection etc. in the cities of England during the period of Industrial Revolution has drawn the attention of eminent scholars like Marx and Engels in the nineteenth century (Engels: 1973; Marx: 1974). The rural-pushed migrants—the surplus people in latent form which is a part of the agricultural population—worked in the urban activities which are very similar to informal sector activities.

Lenin has explained two types in two distinct areas in one industry, which is very similar to informal/formal dichotomy. While discussing the development of mining industry in Russia, he mentions how replacement of one form of social economy by another is particularly clear (Lenin: 1977).

While discussing 'capitalist corporation', Lenin has also brought a distinction between the 'capitalist workshop' and the 'workshop of the small industrialist', which is again very similar to informal/formal dichotomy. The distinction between these two groups of workshops lies in the number of workers simultaneously employed (Ibid.). The organisation of production of large-scale and simultaneous employment of many workers require the accumulation of a fairly large capital. The size of this capital, according to Lenin, determines the form in which the owner himself take part in the establishment.

Marx and Lenin

It is again Marx and Lenin who dealt with the linkages in detail. Lenin, while discussing the forms of merchant capital in small industries has dealt with the aspect of linkages between small and big units analogous to informal and formal sector linkages.

"Having cut off the small industrialist from the finished goods market, the buyer-up now cuts him off from the raw materials market, and thereby brings him completely under his sway. It is only one step from this form to that higher form of merchants' capital under which the buyer-up directly hands out materials to the 'handicraftmen' to be worked up for a definite payment'' (Lenin: 1977, p. 369).

The handicraftman becomes a *de facto* wage worker at home for the capitalist, the merchants' capital for the buyer-up is transferred into industrial capital. The same type of linkages is very much prevalent in *biri* manufacturing in India. Each roller of *biri* is an example of capitalist domestic industry as mentioned by Lenin. The features of capitalist domestic industry show how the small units are linked to large units, similar to informal sector's linkage to FS.

Capitalist domestic industry refers to processing at home, for payment by the piece, of the raw materials obtained from an entrepreneur (Lenin: 1977). The features of capitalist domestic industry are:

(i) multitude of middlemen between the capitalist and the worker. That is, there is a 'sweating system'

(Marx: 1974). Piece-wage facilitates the interposition of parasites between the capitalist and wage labourers, the 'sub-letting of labour'. Given piece wage, it is naturally to personal interest of the labourer to strain his labour-power as intensely as possible; this enables the capitalist manufacturers to raise more easily the normal degree of intensity of labour. This system is similar to sub-contracting system in the present time,

(ii) the truck system: payment in provisions,
(iii) insanitary working conditions: infection and occupational diseases,
(iv) long working day,
(v) drawing of women and children into production, and
(vi) isolation of home workers.

The 'Big Push' and Linkages

According to the theory of 'Big Push', the development process is a series of discontinuous jumps. The functional relationship among the casual factors in economic growth are full of lumps and discontinuities. So, it is argued that a 'big push' or 'minimum effort' is needed to overcome the original inertia of the stagnant economy of a developing country. The arguments are based on the ideas of external economies.

In the early statement of this theory, it was stated that, the effects of proceeding bit by bit will not be equal to sum total of the single bits, due to different kind of indivisibilities, viz., (i) indivisibilities in the production function, (ii) indivisibilities of demand, and (iii) indivisibilities in the supply of savings. So, it is argued, a minimum amount of investment is necessary though not sufficient.

(1) Balanced Growth

In a later stage, Nurkse gave his minimum effort thesis by advocating balanced growth; a frontal attack, a wave of capital investments in a number of different industries. According to Nurkse, the vicious circle of poverty implies a circular constellation of forces tending to act and react upon one another in such a way as to keep a poor country in a state of poverty. This type of situation is summed up as 'a country is poor because it is poor' (Nurkse: 1966). The most important circular relationship of this kind are those that afflict the accumulation of capital. To Nurkse, there is a way out of this vicious circle.

The difficulty caused by small size of the market disappears when there is a more or less synchronised application of capital to a wide range of different industries. People working more productively with more capital and improved techniques in a number of complementary projects become each others' customers. Therefore, there is an overall enlargement of the market which can help to fill the vacuum in the economy of the developing countries.

But the balanced growth doctrine underplays supply limitations. It also does not explain how the balanced growth of different industries reached by an adjustments of demands should be fitted into the type of comprehensive planning adopted by almost all developing countries. Balanced growth involves imposition of new sector on stagnant sector but development implies the process of change of one type to another more advanced type of economy. So, Hirschman argues, the doctrine of balanced growth fails as a theory of development.

(2) Unbalanced Growth

To Hirschman, the best way to achieve economic growth is deliberate unbalancing of the economy in accordance with a pre-designed strategy (Hirschman: 1958). The development will proceed, Hirschman advocates, if we undertake a big push in strategically selected industries or sectors of the economy. The growth will be communicated from the leading sectors to the followers, from one industry to another, and from one firm to another.

Hirschman analyses the concept of unbalanced growth with respect to the relationship between directly productive activities (DPA) and social overhead capital (SOC) by using a production function diagram. He assumes that since the underdeveloped countries have limited ability to utilise resources; SOC and DPA cannot be expanded simultaneously. The problem of the planning is to determine the sequence of expansion that will maximise induced decision making. A country can start either by expanding SOC or by expanding DPA. Either method of unbalanced growth results in additional investment and output. Because of incentives and pressures, the rate of growth is likely to be faster with chronic imbalance.

Any particular investment project may have both: (i) forward linkage, and (ii) backward linkage. The forward linkage encourages investment in subsequent stages of production. Every activity that does not by its nature caters exclusively to find demands will induce attempts to utilise its output as inputs in some new activities. This is known as output utilisation or forward linkage effect. On the other hand, the backward linkage encourages investment in earlier stages of production. Every non-primary economic activity will induce attempts to supply through domestic-production, the inputs needed in that activity. This is known as input provision, derived demand or backward linkage effects.

Satellite industries can be established through backward and forward linkages. The project which has the greatest total linkages should be selected. To Hirschman, the iron and steel industry has the highest combined linkage score.

Theoretical Models

In recent past, various theoretical models have been formulated to explain informal-formal sector linkages. To identify the linkage effects, International Labour Organisation (ILO) has used a four sector model of the economy (ILO: 1972):

$$X_2 = a_{21}X_1 + a_{22}X_2 + a_{23}X_2 + a_{24}X_4 + X_2F$$

X_1= output of small holder agriculture
X_2= output of informal (non-agricultural) sector
X_3= output of private formal sector
X_4= output of the government sector
X_2F = final demand for informal sector output
a = input use co-efficient

By solving input-output row of informal (non-agricultural) sector we can find:

$$X_2 = \frac{a_{21}X_1 + a_{23}X_3 + a_{24}X_4 + X_2F}{(1 - a_{22})}$$

A similar four-sector model based on L.G. Reynold's model is mentioned by John Weeks while explaining the determinants of growth of informal sector (Weeks: 1975). The gross output of IS is exhausted according to the following identity:

$$X_2 = a_{21} X_1 + a_{22} X_2 + a_{23} X_3 + a_{24} X_4 + R$$

Presenting a theoretical framework of an economy which has an IS, one model shows that IS can expand in terms of output and employment even when FS contracts (Chaudhuri : 1989).

There are few other models which try to establish linkages between the incomes of the two sectors. According to Richardson's model, the income of IS is proportionate to formal sector's income (Richardson: 1984). This implies that activities in IS and FS rise and fall together with the growth in IS being dependent on the growth of FS. According to this model, if the demand for IS goods and services is a linear and homogeneous function of urban income, then:

$$Y_i = a\, Y_u$$

Where

Y_i= informal sector income

Y_u= total urban income

Y_f= formal sector income

a= constant

Assuming a variant of economic base theory where total urban income is determined by the level of economic activity in FS; and then by solving equations, he comes to conclusion that:

$$Y_i = \frac{a}{1 - a} Y_f$$

Another theoretical model shows how the level of income in IS is determined by the level of export in FS (Hilhorst: 1984). The simple model is :

$$Y_f = A_f + E_f + E_a - M_a - M_f$$

By solving the equation, one can find the income of IS : Y_i. The model concludes that the income in IS will not only be reduced by increasing imports but also when formal sector's propensity to import from the rest of the world goes up.

A linkage is also established between informal and formal sectors for income tax and minimum wage legislation through diagrammatic representation (Roemer: 1986).

Three Approaches

In general, there are three approaches to the relationship between formal and informal sectors (Tokeman: 1978).

(1) The Benign Relationship Approach

This approach assumes that the relationship of IS and the rest of the economy are benign and complementary. The growth of IS depends on whether it is autonomous or integrated.

(a) Under Autonomy Conditions—the Duality Approach: Under this approach, IS is treated as an autonomous sector having no link with the rest of the economy.

(b) Under Integrated Condition—the Complementary Approach: When IS is integrated to the rest of the economy, the growth potential of this sector is reinforced. So, the linkages are very high, benign and complementary.

Thus, the growth of IS, whether autonomous or integrated, can be evolutionary (Ibid.).

(2) The Subordination Approach

Under this approach, IS is treated as subordinate or dependent and incapable for accumulation.

(a) Under Autonomy Conditions—the Marginality Approach: IS is assumed to be autonomous from the rest of the economy. Subordination of IS operates not through the extraction of economic surpluses but through: (i) lack of access or restricted access to modern inputs and product market and, (ii) existence of surplus labour in both the sectors.

(b) Under Integrated Conditions—the Exploitation Approach: Sometimes, IS is integrated to the rest of the economy in such a way that surplus produced, if any, is extracted from it. Market subordination and lack of access to basic resources are reinforced by several mechanisms, viz., (i) higher price paid for its inputs, and (ii) lower prices obtained for its output. Then, the difference is reaped by FS.

In both the cases, the process of growth of IS will be involutionary (Ibid.).

(3) Heterogeneous Subordination Approach—Neither Autonomous nor Integrated

Under this approach, IS is assumed to be neither as a completely integrated nor as an autonomous sector (Tokeman: 1978). It is seen as a sector which has a significant link with the rest of the economy and simultaneously presents a considerable degree of self-containment. So, this approach has common features of the above two approaches.

Thus, there is a heterogeneous subordination which implies a subordinated relationship for the sector as a whole but resulting from different processes occurring within it. The sub-sectors in IS which are operating under oligopolistic conditions cannot be expected to expand in a permanent way. This approach expects that IS's share of income will decline but IS will not disappear. For this and expected expansion of labour supply for IS, involutionary growth will occur, if no measures are taken.

Types of Linkages

The linkages between informal and formal sectors are not homogeneous and there is a good deal of variation. There is also change in the form of linkages along the process of economic development. In the earlier stage of economic development, linkages are mainly undirectional; at certain level of development, linkages are mutual; and in the final stage, two sectors are more or less integrated.

Broadly, linkages are classified as: (a) forward linkages and (b) backward linkages. Linkages can be classified as: (i) input linkages, (ii) output linkages, (iii) market linkages, (iv) technological linkages, and (v) sub-contracting linkages (Watanabe: 1983).

(1) When IS purchases its inputs, raw materials and machinery from FS, input linkage is established. (2) The linkages of IS through sale of output to FS which uses it is known as output linkages. (3) Sometimes FS sells certain goods after purchasing it from IFS. This type of relationship in marketing the product is known as market linkage. (4) Technological linkages are established in relation to transfer of technology and skill. The technological linkages between IS and FS are expected to take one or more of the following forms (Papola and Mathur: 1983): (a) transfer of equipment, (b) transfer of skill, and (c) sharing of technical know-how. This may take place through (i) sub-contracting arrangement, (ii) sale or hire of machinery, (iii) secondment of skilled manpower from IFS to IS, (iv) organisation of training programme by FS units for the benefit of IS participants, and (v) movement of an employee from FS to start his own enterprise in IS (Ibid.).

The extent and pattern of technological linkages between the two sectors depend on their market relationship such as: (i) independent, (ii) competitive or (iii) interdependent. First, inter-sectoral technological cooperation takes place exclusively in connection with input linkages, when there is an independent product market. Second, if there is competition between the two sectors, IS units will be obliged to use similar technologies used by FS units. But IS units are not likely to get such technologies from FS. Third, when two sectors are interdependent and when IS produces FS's intermediate products, technical linkages between the two sectors are inevitable.

It is observed that inter-sectoral market and technological linkages can serve as important element in the process of industrial development in various ways. First, it provides an assured market for IS products and assured supply of intermediate products for FS. Second, it helps in division of labour and specialisation and, provides opportunities for capital savings and economies in research and development. Third, it establishes technological integration among the two sectors through transfer of technology and skills from FS to IS.

(5) Sub-contracting is a special type of linkage through input or output or technology. Sub-contracting is a business practice whereby the party offering the sub-contract (parent firm, enterprise or company) request another enterprise (sub-contractor, ancillary units or industry) to manufacture or process the whole or part of the product the first party sells as its own (Watanabe: 1983). Thus, sub-contracting differs from the mere purchase of ready-made parts and components. In sub-contracting, there is an actual contract, written or verbal, between the two parties setting out the significance of the order.

There are different types of sub-contracting. First, sub-contracting may broadly be classified as: (i) commercial sub-contracting in which the parent firm can be a wholesaler or retailer who does not undertake any manufacturing or processing of the product itself; and (ii) industrial sub-contracting under which a manufacturer uses the purchased product as part or component of its product (Ibid.).

Second, the forms of sub-contracting can also be distinguished as follows: (i) component sub-contracting in which the parent firm usually concentrates in a limited range of technology intensive segment of the final product and brings other components from sub-contractors; (ii) activity sub-contracting under which the original firm having an integrated plant could have one or more activities undertaken by sub-contractors and then sell the final product under its brand name; (iii) assembly sub-contracting in which the operation of assembling of components to produce the final product, mostly in electronic industry, is farmed out to small and IS units; and (iv) product sub-contracting under which sub-contractors produce the complete product and the parent firm essentially performs the marketing function (Nagaraj: 1984).

Third, sub-contracting can also be classified in another way. FS units can enter into two types of sub-contracting arrangements with IS units, viz., (i) on the basis of prior order without supplying raw materials; and (ii) contracting out certain process with supply of raw materials (Papola and Mathur: 1983).

Fourth, sometimes, sub-contracting is classified into two groups on the basis of dependency, viz., (i) 'tied' sub-contract when the sub-contractor is fully dependent on his parent firm for his market; and (ii) 'untied' sub-contract when the sub-contractor is not fully dependent on the parent firm. It is observed that there are some problems in case of tied sub-contract. A tied sub-contractor's position is more unstable in recessions and more vulnerable to abuses by parent firms compared to untied sub-contractors. Moreover, sometimes parent firms compel the tied sub-contractors to reduce the price of their products.

The objectives of FS units for sub-contracting are many (J. Harriss: 1982). They are : (i) to reduce labour cost by relying on small and IS units for labour-intensive work; (ii) to have control over labour process; (iii) to reduce the level of taxation; (iv) to save capital; (v) to diversify the product range without sacrificing economies of scale; (vi) to increase adaptability to market fluctuation; and (vii) to avoid various labour and industrial laws.

Besides the above objectives, the following factors are responsible for the growth of sub-contracting linkages (Nagaraj: 1984): (i) principle of division of labour, (ii) role of government policies in encouraging ancillarisation, etc., (iii) nature of production technology, (iv) the level of industrial development, (v) historical background of modern industrialisation, (vi) wage differential between IS and FS, and (vii) technological innovation, inter-farm competition and industrial growth.

Mainly, sub-contracting is largely guided by two considerations: (i) lower labour and overhead cost in IS, and (ii) restricting in-house employment of workers to minimise challenge from trade unions.

It is argued that sub-contracting linkage helps industrial efficiency, stimulates industrial activity, contributes to industrial progress, helps in solving the problems of unemployment and poverty, and encourages would-be small-scale industries/IS entrepreneurs to enter into industrial activity.

On the other hand, this system may create the situation for exploitation of sub-contractors by parent firms and may reduce the independence of IS units thereby converting the self-employed into disguised workers.

Constraints on Linkages

There are various constraints on the development of linkages between IS and FS. First, the comparatively high wages and prerequisites of FS employees guaranteed by government policy discourage the movement of skilled workers from FS to IS. Second, in most of the cases, the transfer of better capital and machinery from FS to IS to develop the latter is incompatible with the educational and training level of workers in IS.

Third, the scope of transfer of second-hand machines from FS to IS for reduction in the cost of technological development among IS units is limited because of (i) early stage of development in most developing countries, and (ii) possible adverse effects on quality of product. Fourth, the apprenticeship system has certain limitations such as (i) insufficient inflow of innovative masters/trainers, and (ii) the danger of exploitation of apprentices by the master capitalist in FS.

Realities of Linkages

The linkages between IS and FS have been expressed in different ways. Some view the relationship as benign, the growth of IS being closely linked on its relationship with FS; while others find that relationship is exploitative; the subordination, dependence and exploitation of IS by FS operates either through the lack of resources or through the extraction of surplus generated in it.

In fact, in some cases, the relationship between the two sectors is complementary. FS is dependent on IS for cheap wage-goods, services, cheap labourers, inputs at lower prices and sale of its inputs and; IS is dependent on FS for some of its inputs and sometimes market. Sometimes, IS helps in saving government expenditure; as despite their low economic and social status, waste pickers and collectors in the metropolis of Delhi are making important contribution to society by saving the city government's expenditure for disposing waste (Hayami, Dikshit and Mishra: 2006). The co-existence of capitalist enterprises and the petty producers are beneficial to both sides (Choudhury: 2000). But under integrated conditions, in most of the cases, particularly in case of sub-contracting, IS is exploited by FS through latter's subordination, under the mechanism of which FS appropriates the surplus generated in IS.

Moreover, when in competition with FS, IS is at disadvantage position. But FS does not compete with IS in all the fields. It leaves certain production processes which are not yet profitable to petty producers and IS units.

In fact, IS is no way an independent and exclusive sector. It is linked to FS and the economy through a variety of linkages.

This section analyses the findings from various empirical studies in and outside India to see what are the realities of linkages between IS units/small-scale units and FS establishments.

Marketing System and Exploitation

The modern IS is exploited by FS through marketing system, as was found in Calcutta (Bose: 1978; Romatet: 1983). FS was purchasing goods from IS units at a lower price and was selling it under its brand name at higher price. The IS units were also exploited while purchasing inputs. There was predominance of market linkages in the form of sale and purchase of each other's products between two sectors, though the degree of market linkages differed in metal engineering industry in Kanpur (Papola and Mathur: 1983). Market linkages were not always accompanied by technological linkages but technological linkages accompanied market linkages in most cases. The market linkages acted as a growth-facilitating rather than growth-inducing factors. The efficiency of linked smaller units was much lower than that of their non-linked units in terms of output and value added per unit of fixed capital in Kanpur's metal engineering industry (Ibid.).

Sub-Contracting

The main aim of big and FS units to go for sub-contracting is to save capital and labour cost. The study of engineering industry in Coimbatore (J. Harriss: 1982) show how big units encouraged their skilled workers to set up more ancillary units by selling them second-hand machines at book value and by giving them orders for works to avert labour problems and to reduce wage cost. Among the registered workshops, the provision of specialised works with associated capital saving for the 'parent' firms may be a more important factor than that for labour cost for sub-contracting by large firms. One of the main factors motivating large and FS units in Kanpur metal engineering industry to sub-contract was the minimisation of fixed overhead cost where the demand for the product was of seasonal nature (Papola and Mathur: 1983).

Element of exploitation in sub-contracting is found to exist in industrial sector in India. Exploitation of small-scale units through sub-contracting was confirmed, as in Calcutta, by the study of engineering industry in Coimbatore (J. Harriss: 1982). Although, the element of exploitation of cheap labour in sub-contracting exists, it was far from an adequate explanation of growth of this relationship between large and small firms in Indian industry (Nagaraj: 1984).

There is problem of credit for sub-contractors. In Coimbatore, most of the sub-contractors and owners of firms supplying parts of finished goods to formal and large-scale industry reported cash-flow problems, as they have to give extended credit to the firms they supply, whilst often having to pay on delivery for raw materials themselves (J. Harriss: 1982). They faced credit constraints in these circumstances. They were also subjected to raw material cost squeezes. The parent firms seemed to provide some sort of technical guidance but very little financial assistance (BCCI: 1979; Nagaraj: 1984). On the contrary, sub-contractors extended trade credit to parent firms.

The history of Japanese industrial development presents two industrial organisational forms, viz., (i) external subcontract, and (ii) internal sub-contract (Annavajhula: 1989). As the firms become smaller in size, their involvement in re-subcontracting rises. Various constraints were observed in inter-sectoral linkages and industrial sub-contracting system in automobile industry in Philippines and Mexico (Watanabe: 1983).

Technological Linkages

In Kanpur metal engineering industry, technological linkages either consisted of purchase of some second-hand equipment of large units by small units or supply of specifications of the products ordered by large units (Papola and Mathur: 1983). There were not any significant technology and know-how transfer. Due to movement of skilled workers from large to small units, there was transfer of skill. In metal engineering and chemical industries in Bombay, there were sub-contracting system (BCCI: 1979). The majority of works done by small-scale units were of labour intensive involving simple technology. Some kind of technological guidance was offered by most of the parent firms.

There was transfer of technology from FS to IS of industries in Ghana, Japan, cotton weaving industry in Egypt and, rural machinery industry in China, as are observed by various studies (Hakam: 1983; Karanshawy and Sarkar: 1983; Sigurdson: 1983; Watanabe: 1983).

Brought out components, Ancillarisation and Supplies

The tractor industry depended on sub-contractors for more than three-fourths of its components which help it to reduce the cost of its product (Nagaraj: 1984). The level of brought out components among the producers was also high, according to a study on truck manufacturing industry in India (Lal: 1980). The growth of sub-contracting for components was caused by (i) advantage of the low cost of production of sub-contractors, and (ii) the policy of reservation of certain items in small-scale units.

But, cost reduction was achieved mainly by exploitation of cheap labour in small-scale units and not through the principle of division of labour and specialisation (NSIC: undated). The parent firms provided very little assistance to the ancillaries, as ancillarisation was weakly developed along Thana-Belapur belt (Shaw: 1990). Most of the small units in Jamshedpur manufactured against orders from large units and they did not produce for the market (Gupta: 1980).

Between the mid-seventies and mid-eighties, the issue of scalar linkages in Indian industries was marked by significant research interest. Some useful findings are: (i) small and large-scale sectors have grown in harmony with each other; (ii) production in SSI sector is complementary to large-scale sector; and (iii) direct inter-farm linkages and subcontracting or ancillarisation have remain limited, etc (Shah: 2001).

General Inter-dependence and Linkages

In various wider studies on urban IS, mention has been made on linkages between FS and IS. IS could not develop independently of FS expansion in Ahmedabad (Papola: 1981); while the forward linkages of IS in Nangal was very poor but backward linkages with FS were fairly strong (Dhesi and Wadhwa: 1980). In the waste recycling industry in Bangalore, some workers came from FS bringing in technical skill and managerial expertise, though there was no transfer of production technology from FS (Aziz: 1984).

In Sambalpur (Orissa), while forward linkages between two sectors were negligible, the backward linkages—both direct and indirect—were significant (Samal: 1986, 1990a). There was a greater dependence of IS on FS for their inputs/raw materials/ commodities used or traded but the former's dependence for selling its products was negligible. But, even in limited cases, there was stronger dependence of IS on FS through backward linkages but element of exploitation was not so prominent.

Some scholars (Davies: 1979; Moser: 1978; Remy: 1982) who make a distinction between FS and IS mainly on the basis of mode of production conclude that the capitalist sector/FS develops owing to the benefits derived by it from IS/petty producers.

According to the studies conducted by the International Labour Organisation (ILO) in cities and towns in Asia, Africa and Latin America, the forward linkages of IS with FS was virtually absent (Sethuraman: 1981). The IS in these cities and towns sold output to households and individuals. Similarly, there were little backward linkages between the two sectors in these cities and towns.

One study on Durgapur (Mukherjee: 2004) concludes that appropriate policies for the optimum development of IS include, among others, well-coordinated formal-informal linkage. Informal manufacturing sector can play a very vital role in shaping the development of a region through its diverse interaction and complementarities with formal organised sector (Ibid.).

Summary

From these studies mentioned above, it is observed that there were more linkages between the two sectors through sub-contracting in the engineering as well as automobile industries. The engineering and auto industries greatly depended on sub-contractors (SSI/IS units) for their components. In certain cases, they took initiative for ancillary development and for sponsoring establishment of new sub-contractors mainly to reduce cost of production by getting components at cheap price from sub-contactors which, in turn, led to exploitation of labour in SSI/IS units by paying them low wages. Of course, parent firms sub-contracted components requiring only labour-intensive methods.

Though, the small and IS units received technical guidance from the parent firms, in most of the cases, technology transfer was very limited. The technical linkage was either in the form of purchase of second-hand machine or getting specification of product but did not involve any significant technology and know-how transfer.

It is also found that, as the industrial development proceeds, there is transformation of linkages and sub-contracting arrangements including re-subcontracting chain.

Findings from Orissa

We have conducted two major studies in Orissa relating to linkages. One study (Samal: 1990e) deals with linkages between informal and formal manufacturing sectors in a planned industrial city (Rourkela). The other (Samal: 1994b, 1995, 1996) is on the linkages of Nalco Angul sector with the manufacturing sector in Orissa. The former has studied the linkages between IS and FS inside a city, while the other has dealt with linkages of a single FS unit with both FS and IS manufacturing units in whole of Orissa.

Linkages in a City

The study in Rourkela classified the linkages of IS with FS on the basis of its inputs and outputs. The relationship through its output was termed as forward linkages and via its inputs as backward linkages. The study emphasised on identifying all these types of linkages of IS in details: (i) forward linkages through (a) output sale, (b) sub-contracting, and (c) marketing of products; (ii) backward linkages through (a) purchase of inputs and raw materials, (b) acquisition of skill and technology, and (c) procurement of credit; and (iii) technological and institutional linkages.

The major objectives of the study in Rourkela were: (i) to measure the extent and portray the structure of the modern informal manufacturing sector (IMS) in the city, (ii) to examine the linkages between IS and FS in the manufacturing activity in the functioning of IS, and (iii) to study, in particular, the linkages of IMS units with Rourkela Steel Plant (RSP) in public sector and its impact on them.

We collected the information about the universe of modern IMS (that is total number of units registered under Shops and Commercial Establishment Act) from the office of the Assistance Labour Commissioner, Rourkela, and from these units, we sorted out the modern IS units in manufacturing activities. The universe in modern IMS was divided into different sub-groups on the basis of main inputs/raw materials used for production and from each sub-group, some units were selected by random sampling. At least, a minimum of 25 per cent of the total units was taken as our sample for the survey.

While analysing data, these units, wherever necessary were further classified under the following categories: (i) those which were treated as ancillary to FS units, (ii) those who did not get ancillary status but dependend mostly on order from other units, (iii) those who came up due to availability of raw materials, inputs from other units, (iv) those which served to the changing demands of local population due to rapid urbanisation consequent upon the establishment of FS units, and (v) those who come up due to locational and infrastructural advantages offered near big FS industries but have no interdependence.

We administered structured questionnaire directly through personal interviews to the IMS entrepreneurs in our sample to get detail information about the working and the linkages of enterprises. The size criteria was adopted by us to demarcate IS units from FS units. The total effective sample in IMS was 54 which was distributed among seven broad groups (viz. mechanical spare parts-15, repair and garage-7, furniture-5, fabrication-5, printing press-5, food grain-based-2, others-5, non-linked units-10). We collected information from FS manufacturing establishments (both public and private) including RSP through another structured questionnaire. The total effective sample in FMS in Rourkela region was eight including RSP in public sector and Larsen and Toubro in the private sector.

The western part of Orissa particularly from Bargarh to Birmitrapur covering undivided Sundargarh and Sambalpur districts is an industrial belt in which large number of big, medium and small-scale industries are found. The location of our earlier study on the whole aspect of urban IS including linkages was Sambalpur town which is also in the western part of Orissa (Samal: 1986, 1990a, 1990b). Sambalpur is the district and divisional headquarters and, is primarily an administrative and commercial town.

Our study of urban IS in Sambalpur shows that the direct forward linkages of IS with FS was limited. A large proportion (more than three-fourths) of sale was made directly to consumers and households. The sub-contracting arrangements which was a tied sub-contract was found only in case of *biri* units in manufacturing which neither helps in industrial development nor in transfer of technology. In case of regular arrangements for sale of output, the difference between selling price and market price was reaped by other comparatively bigger IS units particularly trading units.

The IS units had backward linkages through input, credit and partly through acquisition of skill and technology. About one-third of the total sample units had direct backward linkages fully or partly with FS enterprises (which are mostly outside the city and the state) for their inputs/raw materials/ commodities used or traded. There was also indirect backward linkages of IS with FS through retailers and wholesalers.

In that study (Ibid.), around one-third of the sample units received loans from the nationalised commercial banks and government agencies. The linkages of IS establishment with government and other public system by way of regulation and, input and utility supplies were not also insignificant. Transfer of technology took place to some extent from FMS to IMS through sale of machinery but the proportion of workers acquiring skill from FS was very negligible.

The study in Sambalpur (Ibid.) concluded that in the development process, some sub-sectors which are linked to industrial development will grow at an increasing rate and others which face stiff competition from FS will grow at a decreasing rate. Though there may be competition between the two sectors in certain activities, one sub-sector in FS may be complementary to a different sub-sector in IS. In certain other activities, IS is able to continue in its operation for its personalised services, in spite of stiff competition from FS. In most other activities, it is more likely the growth of FS and IS gets linked together.

Though, we came to the above findings and conclusions in our study (Ibid.), we were not satisfied with our survey results particularly with respect to linkages, since there were some limitations. We studied IS of Sambalpur town as a whole consisting of different sub-sectors (i) manufacturing, (ii) trade which included: (a) wholesale, (b) retail, (c) hawking and (d) hotel, (iii) transport, (iv) services, (v) agriculture and allied, and (vi) construction. The above generalisation relating to linkages applies to all sub-sectors of IS as a whole. But to our disappointment, we did not find any noticeable linkages between IMS with existing FMS in the town via either their inputs or outputs; though IMS was an important sub-sector in IS. Perhaps, it is because, Sambalpur itself is not an industrial town. It is mainly an administrative and commercial town. Whatever linkages of IS was with FS in other towns and cities mostly outside the state.

For this reason, we have undertaken another study on linkages between IMS and FMS in Rourkela which is in the same western part of Orissa and in the same industrial belt. Rourkela is an industrial town with a big public sector steel plant. The steel plant is believed to have greatest linkage effects. So, Rourkela, where RSP is operating, is most suitable town for the study of linkages between modern IMS with FMS. Rourkela falls on Howarah-Bombay section of South-Eastern Railway and is 430 km from Calcutta. It is situated in Panposh sub-division of Sundargarh district in Orissa. It is bound by the rivers Koel and Brahmani on the northern and western sides respectively. It is, at present, a modern planned town which was a small tribal village six decades ago. Its population, according to 2001 Census, is about 4,84,292. The township is spread over an area of 157.17 sq. km.

Sundargarh district in which Rourkela is situated was carved out in the map of Orissa after the merger of two ex-Princely states, Gangpur and Bonai in 1948. The total population of the district as per 2001 Census is around 18,29,412. A sizeable portion of the important minerals of Orissa are available in abundant quantities in the district, such as iron ore, manganese ore, limestone and dolomite which are basic and important raw materials for steel production. The establishment of RSP at Rourkela was possible for the vast deposit of these minerals in Sundargarh district.

RSP is the first integrated public sector steel plant in India set up in mid-1950s in collaboration of the earstwhile FRG to meet the growing demand of steel as well as to lay down a strong industrial base for the economy. The raw material requirements of RSP are mainly met from its own captive mines. The growth of SSIs are stimulated in the town by RSP and other big industrial units like L&T in and around Rourkela. Most of the SSIs utilise the products/by-products of RSP for their production to serve the consumer market. They also supply the consumables, spares and raw materials to RSP. RSP has been constantly increasing the placement of orders on SSI units and offering technical guidance and material assistance.

There has been shift in occupation of both tribals and non-tribals from agriculture and allied activities to services in industries in both urban and rural areas around the city after the establishment of RSP and other big industrial establishments in Rourkela region. The level of income of the people in the region have increased manifold owing to industrialisation (Mohapatra: 1990). Industrialisation benefited the people in the area in various ways by providing direct employment and encouraging the villagers to produce agricultural products for urban market. But industrialisation is not an unmixed blessing since it has led to various social and environmental evils as well as to rising cost of living. By and large, RSP and industrialisation has brought a transformation in the economic, social, political, cultural and environmental aspects of the area (Ibid.).

Structure of Linkages: There was little difference between linked and non-linked units in IMS regarding registration, ownership, structure, premise and period of operation. But great difference between these two groups was observed in relation to fixed capital and assets, working capital and goods produced (Samal: 1990e). The linked units were comparatively more capital-intensive; and goods produced and working capital per unit was also significantly higher in these establishments compared to non-linked units in IMS. The entrepreneurs in the former were comparatively older and educated than those in the latter. The capital investment and education of the workers in IMS were most important determining factors in establishing linkages with FMS (Ibid.).

Pattern of Linkages: Most of the linked establishments in IMS had linkages with RSP and around half of their sales went to RSP (Ibid.). The linked IMS units had direct linkages with RSP either through (i) selling their outputs to RSP, or (ii) purchasing their inputs from RSP. On the other hand, half of the sample FMS units had linkages with IMS. Comparatively bigger FS establishments including RSP and L&T had established linkages with IMS in Rourkela. But most of them were purchasing spare parts and consumables from the latter. Some FMS units had linkages with other comparatively bigger units. Only few supply IS goods after procuring and processing it. About three-fourths of non-RSP FS units had linkages with RSP. They were selling goods which were procured from IS and processed in their units. Thus, RSP plays an important role in establishing linkages directly and indirectly with IS units and some of the IMS units were sub-contractors' sub-contractors mostly of RSP.

While the factors motivating IMS in Rourkela to establish some sort of linkages with FMS were (i) to utilise productive capacity economically, (ii) to protect the establishments against competition, (iii) to have an assured market for their product, or (iv) for stability; the motives of FMS to sub-contract to the informal units were (i) for reduction in production cost, and (ii) due to lack of complementing equipments (Ibid.).

Half of the sample IMS units depended on FS establishments for their raw materials and inputs. It is to be mentioned that the registered small-scale units were getting their quota of raw materials from the organised agencies of the Government of Orissa and RSP. But IMS units were procuring their required inputs from scrap dealers, hoarders and black marketers. Thus, output of IMS units was not used as input by our sample FMS. The FS units purchased mostly consumables and spare parts from the informal establishments (Ibid.).

In case of regular arrangement, only in few cases, the price offered by FS was lower than the market price. Around two-thirds of IMS units having regular arrangements with FS were controlled by the latter. In these circumstances, there was credit transaction in some cases but credit varies from one to three months only, without payment of any interest by FS units to IS units.

Almost all the IS entrepreneurs felt that their linkages with FMS was beneficial. Some of the linked informal units reported growth in their sale and employment due to establishment of their linkages with FMS. Though the growth in employment was not so much prominent, the growth in sale was significant in IMS having linkages with FS. Mainly the gains from linkages are for marketing their products. Similarly, three-fourths of the sample FMS including RSP and L&T were satisfied with the quality of goods supplied by IMS. It was manageable to those who were indifferent about the quality of the goods supplied by IS units (Ibid.).

The main problem faced by a handful of the linked IMS units was due to irrational order from FMS. On the other hand, the main complain of a formal sector unit—the public sector giant RSP—was that these linked informal units took order beyond their capacity. But, in general, most of the FMS units including L&T had not faced any problem due to their linkages with IMS. Maybe, their main concern was for reduction of production cost and avoidance of labour problem (Ibid.).

In reality, for any materials or job works, RSP floats tender and hence there is keen competition among small and informal units to get the contract. They bribe to the Purchase Department of RSP to get the contract and also bribe to the Quality Control Department very often to get their sub-contract and job works okayed. So, in this case, it is difficult to find out who is the real exploiter since there is an unholy alliance between the two. In reality, it is not the entrepreneurs of informal manufacturing sector who are exploited but the workers in IMS who are exploited under this type of mechanism. The competition among small SSI and IMS units to get the order to supply raw materials or get the job contract at a low cost under a tender system puts IMS in disadvantageous position in general. But this cost as well as the extraction of surplus from IMS by FMS no way hampers the interest of the owners/entrepreneurs of IMS. The whole burden is shifted to IMS workers who are ultimately exploited under this system of linkages and relationship through low wages, denial of various facilities and violation of labour laws. Therefore, the greatest victim of informal sector's linkages with FS are the workers and not the entrepreneurs of IMS in Rourkela (Ibid.).

Thus the linkages between the two sectors served the interest of IMS entrepreneurs and some FS executives. The relationship between the two sectors was complementary; though in cases of regular arrangement and subcontracting, element of exploitation existed since FS was in advantageous position due to its strong bargaining power. But it was very difficult to quantify dependency and exploitation of IS by FS in a city dominated by the public sector undertakings.

In the present situation of liberalisation and opening up of the economy along with the dereservation of some items previously produced by small and informal sector will put IS in a most disadvantageous position. Liberalised Industrial Policy in consequence (i) make small-scale and informal sector to face the stiff competition from large business houses and multinationals on the one hand, and (ii) on the other, lead to exploitation of existing linked IS units owing to the decline of public sector since exploitation by public sector formal units is comparitvely less (as found by our study).

In the case of IS units with linkages with FS, a cautious and careful policy of utilising the growth potential as well as regulating the arrangement so as to minimise the exploitative features of the linkages need to be worked out. In view of the exploitative relation, policies for the development of IMS must take the relationship between the two sectors into consideration. Policies may be undertaken to encourage the linkages between the two sectors but not under integrated condition. Moreover, encouragement for linkages with public sector formal units may be given as well as the labour laws should be tightened in IS itself in order to stop exploitation of its workers.

Linkages of a FS Unit in Orissa

In two of our studies relating to NALCO (Samal: 1994b, 1995) we have studied the linkages of NALCO Angul sector with the manufacturing sector in Orissa. The study found that NALCO has given some momentum to trade, transport and economic activities in NALCO Nagar in particular and in Angul area in general. There was a rapid growth of informal activities around NALCO Angul sector, mainly composed of shops and commercial establishments.

However, the direct impact of NALCO Angul sector in manufacturing activities in both the undivided Dhenkanal district and the state of Orissa is very negligible. The forward linkages of NALCO Angul sector with the manufacturing sector of Orissa is primarily established through the sale of aluminium ingots and wires. Around 67 manufacturing units in Orissa were purchasing aluminium metal directly from NALCO and 13 units indirectly through OSIC (Ibid.). Mainly, aluminium consuming units like cable, utensil units have linkages with NALCO Angul sector but dependence of cable units is more on NALCO. Around one-fifth of the total purchase of main inputs and raw materials (in money terms) of the forward-linked manufacturing units was from NALCO Angul sector. But only 3.72 per cent of the total sales turnover of aluminium metal by NALCO was within Orissa (Ibid.). Thus, the forward linkages of NALCO Angul sector with the Orissa's manufacturing sector is negligible.

Power and alumina are the major inputs of NALCO Angul sector which are available from its captive plants. NALCO was purchasing mainly ancillaries, spares and consumables from about 65 units in Orissa (Ibid.). So, the backward linkages of NALCO Angul sector with the manufacturing sector of Orissa were mainly established through purchase of ancillaries, spares, consumables and miscellaneous items and not through major raw materials and inputs. The annual purchase of raw materials, inputs, spares, ancillaries and consumables by NALCO Angul sector was around Rs. 150 crore. The purchase from inside Orissa constituted only between one to three per cent of the total annual purchase (Ibid.). Therefore, the backward linkages of NALCO Angul sector were very negligible. The backward linkages were not only negligible but also erratic and irregular.

Out of the total 26 sample backward-linked units of NALCO Angul sector, only 10 units, all in Dhenkanal district, can be regarded as IS units since their employment size was within 10. Out of this 10 IS units, nine are post-1982 units; three were producing ancillary and spares, and two chemicals for NALCO (Samal: 1994b).

Out of the total 30 sample forward-linked units of NALCO, only six were in informal sector, one in undivided Dhenkanal district and five in undivided Puri and Cuttack districts. Among the IS units, three were producing cables and one utensils. Half of the IS units were pre-1987 units and other half post-1987 units (Ibid.).

NALCO had neither ancillary unit nor any sub-contractor when our survey was conducted during first half of 1993. There was also no technological linkage of NALCO with these units. Though NALCO is using improved and most advanced technology imported from France, the downward transmission of technology to the backward-linked units is nil (Ibid.). Thus, the linkages—both forward and backward—of NALCO Angul sector with the manufacturing sector in Orissa were insignificant.

Ancillary Development Programme

Ancillary Development Programmes of public sector undertakings, state sector units, large- and medium-scale industries in private sector help in the development of SSI and IS units through linkages. The progress of Ancillary Development Programme is not satisfactory in Orissa, though there is scope for development of ancillary industries. Since the early 1970s, Union Government launched ancillarisation of SSI and IS units through its public sector undertakings. The success of ancillarisation programmes in any state, however, depends upon the establishment of PSUs and large- and medium-scale industries in the state. In Orissa, besides more than 12 PSUs, there are about 150 large and medium industries creating a better scope of ancillarisation. In the late 1980s, there was some visible ancillary development programme. Few steps have been taken up by some of the mother plants towards off-loading of their requirements to SSI and IS units. But, it is not quite encouraging. RSP has ancillarised 25 SSI and IS units. NALCO has given ancillary status to only 27 units and HAL (Sunabeda) to three at Sunabeda. Mahanadi Coal Fields Ltd. has also given ancillary status to 55 units (Government. of Orissa: 2001). There is uncertainty among ancillary and small units because of the reluctance of large PSUs to follow guidelines of Bureau of Public Enterprises. So far, the effort of the state government in this direction has not been satisfactory, though Directorate of Industries, Government of Orissa is supposed to coordinate with large and medium industrial units to obtain ancillary status or registered manufacturer/supplier status for small- and medium-scale industrial. There is further scope for ancillary development in Orissa with Paradeep Phosphate Ltd., L&T and other large and medium industries in the state.

Summary

Thus, it is apparent from earlier case studies in and outside India on linkages between IS and FS that in some cases, the relationship between the two sectors was complementary. But under integrated condition, in most of the cases, IS was exploited by FS through latter's subordination under a mechanism whereby FS appropriates the surplus generated in IS. It is observed that there were more linkages between the two sectors through sub-contracting in the engineering as well as automobile industries.

The formal sector establishments never gave any financial assistance to their sub-contractors and SSI/IS units. Rather small and IS units offered extended credit to FS and big units since they did not get prompt payment and hence face cash-flow problems. Technology transfer from FS to IS was very limited. The technical linkage was either in the form of purchase of second-hand machine or getting specification of product but did not involve any significant technology and know-how transfer.

The linkages between IS and FS is heterogeneous. In most of the cases, the forward linkages of IS with FS were negligible but backward linkages—both direct and indirect—were prevalent. However, whatever may be the linkages, always IS was at disadvantage due to its weak bargaining power. In most of the cases, IS was exploited by FS either through input or output linkages.

There were complex set of factors which influence sub-contracting out of which saving labour cost and capital cost was prominent. However, there were various constraints on the development of linkages between the two sectors in certain industries. As the industrial development proceeds, there is transformation of linkages and sub-contracting arrangements including re-sub-contracting chain.

Iron and Steel industry is believed to have the highest total linkages which may not be found in case of aluminium industry. From our studies in Orissa, same trend is observed that the linkages—both forward and backward—of NALCO in Orissa are very negligible in comparison to RSP in Rourkela.

4

Migration

The interrelationship between migration and development is universally recognised. The migration pattern varies according to socio-economic context (Simmons: 1984). The neo-classical model is most frequently used to interpret economic development and the pattern of migration in the context of the developed countries in the stage of their industrialisation. But applicability of this model for most of the developing countries would seem at best to be only partial.

According to the rural development model, the social and economic policies in many developing countries have 'urban bias' which tend to subsidise high technology and capital-intensive industry while keeping food prices low in urban centres. These policies, it is argued, are the major causes of rural poverty and migration of workers from rural to urban areas.

The labour-surplus model argues that, in many countries, there is surplus labour in both rural and urban centres. So, it is claimed that, circulation and return migration are common.

The plantation model, which is a sub-model of the labour surplus approach, is distinguished from the general model because the key element in economic change is not urban industry but rural plantation agriculture, forestry or mining (Ibid.). In most of the cases, mines/plantation/forestry are found somewhat distant from the available pool of labour. So, these activities attract labourers from rural areas. However, in some cases, the labour force demand is subject to short-term fluctuations related to the demand for a raw material in the international market. Each migration pattern will have consequences that are unique to itself and to the economic context which brings it about.

There are also some specific theories to explain migration. Ravenstein, in the 1880s, first proposed his laws of migration, which states that migrants move from areas of low opportunities to areas of high opportunities. Building on Ravenstein's law, Lee (1966) divided the forces exerting an influence on migration perceptions into 'push' and 'pull' factors. The movement of workers from subsistence sector to capitalist sector is vividly explained by Lewis (1954) and is further elaborated by Fei and Ranis (1961). Sjaastad (1962) advanced a theory of migration, which treats the decision to migrate as an investment decision involving an individual's expected cost and returns over time.

Two Leading Theories

The most popular rural-to-urban migration model is that of Todaro (1969). In contrast to earlier models of one-way migration, Todaro regarded the migration as a two-stage phenomenon. He recognised a dichotomy in urban economy while analysing rural-to-urban migration.

According to Todaro model, in the first stage, the unskilled rural workers migrate to an urban area and initially spend on certain period of time in urban traditional sector. In the second stage, they eventually attain jobs in more permanent modern sector. In Todaro's model, the decision to migrate from rural to urban areas is fundamentally related to two principal variables, viz.: (i) the urban-rural real income differentials, and (ii) the probability of obtaining an urban job.

A basic hypothesis of probabilistic migration model of Todaro is that informal sector employment is a temporary staging post for new migrants on their way to formal sector employment. It assumes two permanent sectors in the economy; one is rural sector specialising in production of agricultural goods and the other is modern urban sector. The urban sector is divided by Todaro into two sub-sectors: (i) modern sector (akin to formal sector), and (ii) traditional sector (similar to informal sector) which includes all workers not regularly employed in urban modern sector, that is, overly employed, underemployed, or sporadically employed and those who are employed in petty retail trades and services.

There are four basic features of Todaro model. First, migration is stimulated primarily by rational economic consideration of relative benefits and costs. Second, the decision to migrate depends on 'expected' rather than actual urban-rural wage differential. Third, the probability of obtaining an urban job is inversely related to the urban unemployment rate. Fourth, migration rates in excess of growth rates of urban job opportunity are not only possible but also rational and probable in the face of continued positive urban-rural 'expected' income differential. In Todaro model, 'expected' gains[1] are measured by: (i) the difference in real income between rural and urban job opportunities, and (ii) the probability of a new migrant obtaining an urban job (Todaro: 1976).

By incorporating informal sector into standard expected income Todaro type migration model, Stark postulates a two-pronged planning horizon that rural-to-urban migrants may consider (Stark: 1982a, 1982b). According to Stark's model, many rural-to-urban migrants rationally, though involuntarily, join the ranks of the urban unemployed since there are fewer high-paying formal sector jobs than their numbers in urban centres. Yet, migrants may willingly remain unemployed for a long period in towns as an optimal strategy or investment in search of high-paying FS jobs. During this period, they may not accept an IS job even if it is available at the competitively determined market-clearing wage. They may live in town during this period of job search depending on their own savings, familial or similar support.

The Stark model assumes that migration decision are based on a two-period planning horizon that rural-to-urban migrants may consider either of the two competing strategies: (i) join IS in the first period and, while there, seek FS employment. If FS employment is not secured in the second, IS employment is sought; (ii) search intensively (full time) for a FS opening in the first period. If FS employment is not secured, in the second period, an attempt is made to join IS.

Rural-to-urban migrants may choose to remain unemployed for yet another reason. They may think that their acceptance of low-paying low-skill IS job could be constructed as a signal, albeit an imperfect one, by FS, reflecting their (apparently) inferior labour force qualifications and personal traits (Ibid.).

Types of Migration

There are various types of and bio-polar distinction of migration (Samal: 1984). Besides rural-to-urban migration, there are also other types of migration. One type of classification is between external and internal migration. External migration refers to movements from one country to another country, for instance, the exodus of refugees from Pakistan to India in the 1940s. Internal migration refers to movement within the country, from one state to another state, or from one area to another area. Second, migration can broadly be classified into two categories, viz., inter-state and intra-state. The movement of people crossing state boundaries is known as inter-state migration, while intra-state migration refers to movements within the state. Inter-state migration is relatively small in India mainly due to difference in language. However, there is a large number of inter-state migrants in Maharashtra, West Bengal, Karnataka and Delhi in which the cities like Mumbai, Kolkata, Bangalore and New Delhi are located, drawing migrants from other states.

Third, another type of distinction is between positive and negative migration. Positive migration is attracted by opportunities for economic improvement and negative migration refers to movement of labourers driven away from rural areas by lack of opportunities. Fourth, migration may be classified as temporary or permanent. Permanent migration takes place, when migrants usually go to another place from his native place to settle permanently. The temporary migration to towns take place during floods, droughts, cyclones or other types of natural disasters. Most of them return to their native places after the worst of the natural calamity is over. Fifth, migration can be classified either as rural-to-urban or rural-to-rural migration (Ibid.).

Another type of migration is female migration. A large number of migration of women takes place following their marriage. Most of the rural-to-rural migration is by women in India. This is either marriage migration or associated migration accompanying their migrant husbands. A large proportion of female migration also takes place in India for economic reasons (Ibid.).

The migration caused by a different type of disaster is known as disaster migration. The migration of refugees from Pakistan since 1948 is biggest disaster-induced migration. Migration may also be induced by social disasters. This migration is individually-based and is in the nature of the 'system-avoidance migration'. Many sex workers in Kolkata joined their profession after being deserted by their lovers who brought them to the city from their native places (Ibid.).

One type of migration is known as organised migration, such as migration to newly growing economic centres. Certain types of migration can be grouped under voluntary migration, step migration, seasonal migration, etc.

Seasonal migration has drawn the interest of the scholars in recent times. Migration, particularly of poor, is seasonal, temporary and, most of which remains within rural areas. Though, the daily earnings for migrants at the destination may be only marginally higher, migrants undertake this migratory work to maintain or slightly improve their situation at home (Haan and Rogaly: 2002). Seasonal labour migration is an increasingly important aspect of rural livelihood in tribal areas of Western India. Such migration has to be seen as integral to the coping, survival and livelihood strategies of tribal farming families. The forces leading to migration in these areas are as much to do with the social relations of dependency and indebtedness which subsistence failure entails, as with ecological decline (Mosse, Gupta, Mehta, Shaha and Rees: 2002). Over 5,00,000 people are regularly engaged in seasonal migration for paddy work into southern West Bengal. In West Bengal, seasonal migration can involve practical welfare gains. Though an informal wage floor has been put into place and managed by the peasant union, the costs and risks of migration remain high (Rogaly, Coppard, Rafique, Rana, Sengupta and Biswas: 2002).

However, more research and discussions are going on rural-to-urban migration in connection with urban IS, which has drawn the attention of economists, sociologists, policy makers and politicians.

Determinants of Rural-Urban Migration

Various empirical studies show that migration does occur from low to high income regions but some debate remains regarding the relative importance of the characteristics of origin and destination in explaining (Greenwood: 1975). Differences are observed in the magnitude in which income and other factors influence various population or labour force in their decision regarding whether and where to migrate.

The migration is not an atomistic individual reaction to external forces but that such movements are tightly structured by social institutions, households, kinship and other networks (Haan and Rogaly: 2002). Migration has been caused by and, in turn, influences poverty and livelihoods for men and women and these relationship have changed over time (Haan: 2002).

A number of independent variables used to explain migration include: (i) wage or income levels, (ii) unemployment rate, (iii) the degree of urbanisation, (iv) the distance between origin and destination, (v) the friends and relatives of residents of origins in the destination, and (vi) population in origin and destination (Yap: 1997). Age, education, income, employment status, race and castes are prominent personal characteristics that appear important in influencing migration (Greenwood: 1975). Macro-economic factors like (i) agricultural land scarcity, (ii) urban bias, (iii) fuel cost, (iv) public investment, etc., are set of forces driving migration (Williamson: 1988). However, various empirical studies have given importance on certain other factors determining the movement of people from rural areas to urban centres (Samal and Mishra: 1999).

Education

There are two conflicting views on the role of education influencing migration (Stark: 1982a). One view is that the better educated rural youths who acquire non-rural specific human capital in higher amount are induced to migrate to urban centres since they are less averse to risk. So, promotion of schools in rural areas stimulates out-migration by more selective rural youths. Some of them migrate to improve their education and skill, while others move to towns dissatisfied with the prospects of rural life.

The alternative view is that poor educational opportunities in rural areas encourage rural families particularly from wealthier class to send their children to urban areas with better schools and educational institutions. So, the question arises whether the absence of education or its availability is responsible for generating rural-urban migration.

Contacts and Distance

Without kin contacts, migrants' chance of getting employment is greatly reduced particularly in formal sector (Oberai and Singh: 1983; Rempel: 1996; Yap: 1977). Moreover, the availability of family members in urban centres who can help in getting a job in town may actually facilitate the movement of the people from rural to urban centres. The existing commercial and trade interaction between rural areas and urban centres serves as a source of information about the conditions and opportunities therein, which in turn, helps the movement. The presence of friends and relatives in urban areas who often provide help and financial security and the desire of migrants to break away from the traditional constraints of inhibiting rural social structure have been cited as likely determinants of rural-to-urban migration.

Thus, destination contacts have a positive effect on migration to a specific area, when contacts are measured by the presence of parents, friends and relatives in the city, by potential ethnic contact, by language similarity between the origin and the destination or by the stock of persons in the destination who had migrated earlier from the home location.

Distance, on the other hand, has a deterrent effect (Ibid.). With few exceptions, the distance co-efficient in the migration function is negative and significant. Usually, migrants to urban centres came from the same or neighbouring states or districts.

It is also pointed out by some researchers that the benefits resulting from migration need not be large to justify the direct expenditure made for the migration from the origin to destination (Nelson: 1959; Sjaastad: 1962). There are three important explanations for the importance of distance in explaining the movement of the migrants (Greenwood: 1975). It is argued that migration decreases substantially with increased distance because distance serves as a proxy for (i) the transportation cost of movement, (ii) psychic cost of movement, and (iii) the availability of information.

First, it is hypothesised that as distance increases, direct transport and moving cost also increase. However, these costs are usually small, typically a fraction of the added income required to induce a migrant to move long distance. In many cases, the benefits associated with migration are not particularly high and therefore, direct transportation cost need not be great to discourage migration. The consensus of the researchers on migration appears to be that private benefits of migration are substantial enough to offset most reasonable direct transport expenses (Ibid.). The rate of return on investment in migration may, however, be different to a great extent for various sub-groups of migrants like entrepreneurs, workers, etc. It is also sometimes recognised that the relationship between out-of-pocket money cost and geographical distance are likely to be correlated with the income level of the migrants because holding of physical assets and income are associated.

Second, it is argued that distance undoubtedly reflect the psychic cost associated with migration. Long distances between the origin and destination reduce the chance of frequent visits to the native places. So, the psychic costs involved in migration are substantial and are closely related to distance. Therefore, the notion that migration decreases with increased distance has been attributed to the fact that distance serves as a proxy for psychic cost of migration.

It is, sometimes, further argued that psychic costs are not real resource cost and should not be included as part of the social investment in migration (Sjaastad: 1962). On the other hand, it is pointed out that psychic costs are similar to consumers' surplus for the potential migrants. Nevertheless, psychic costs do influence resource allocation by influencing the private cost of migration. Of course, difficulty arises in different measurement of the psychic costs of migration. Sometimes an alternative to direct measurement of psychic costs is also suggested. Psychic cost can represent permanent transportation cost by quantifying the needed frequency of visit to the place of origins so as to negate the agony of departure from family and friends (Schwartz: 1973). So, it is contended that psychic costs are likely to rise with age since the frequency of visits to natives is likely to increase with age. Therefore, the deterring effects of distance also rise with age.

On the other hand, it is pointed out that the relationship between increased education and psychic cost is non-increasing. So, the deterring effects of distance do not rise with the level of education. It is also found in a study (Ibid.) that while the effect of age on distance elasticity is weak, the increased level of education significantly reduces the deterring effects on migration. The study concludes that the distance effect is really an information effect and not the result of psychic cost.

Third, it is hypothesised that the availability of information concerning alternative destinations plays a prominent role in the potential migrants' decision regarding a destination (Greenwood: 1975). Persons are more likely to migrate to places about which they have atleast some information rather than to areas about which they have little knowledge or nothing. Thus, the factor responsible for the proposition that migration decreases with distance is that information decreases and hence, uncertainty increases with increased distance from the origin of a person. It is also observed that past migration of relatives, friends and native persons is an important determinant of the distribution of present migrants and when account is taken of past migration, the current direct effect of distance are not nearly so great as they otherwise seen (Ibid.).

The fact that the more educated tend to be less deterred by distance than the less educated has also been interpreted as support for the information hypothesis. Less educated rely more on information from friends. So, it is argued that information from friends and relatives decline with distance and thereby deter migration. On the other hand, for more educated, information from mass media including print and electronic media is a more important source. Therefore, information from mass media does not diminish as rapidly with distance, accounting for the longer trips. But this interpretation is ambiguous (Yap: 1977).

Thus, gross migration declines perceptively with increased distance. However, the issue of the relative importance of various economic and non-economic factors for which distance is a proxy remains unresolved.

The role of distance is not so great as assumed by the dominant hypotheses. Distance alone does not determine the decision to migrate in case of workers in rural areas. In Orissa, in past, the labour contractors played a major role for migration of workers to Kolkata to work in the dock, jute mills, to Assam to work as pluckers for tea garden, and to Ahmedabad and Surat for working in textile mills (Samal and Meher: 1998).

Second, workers may like to move to a large-distant place than to a short-distant place, if the transport facilities to the former are better than the latter. Moreover, the type of transport—railway or road—is a factor which enters into the calculation. Workers may prefer a destination which is linked by the railway than a place, may be nearer, but linked by road. The large migration of workers from the coastal belt of Orissa to Kolkata and from Bolangir-Kalahandi region to Sambalpur-Rourkela region in Orissa was possible due to the railway link. Even if there is rail link between the origin and destinations, the destination having a direct link from the nearby railway station is more preferred than the destination to which one has to reach after changing two or three trains. For these reasons, workers usually move to metropolies or capital towns which are well connected with different parts of the country or the state (Ibid.).

Third, a single worker may not like to go alone to a short-distant destination but he may prefer to a large-distant place if he moves in group with his friends, relatives or socio-cultural known persons of his area. They also move to large-distant place when the labour contractors recruit them for destination bearing the expenditure towards transport cost and initial expenditure for food, etc. (Ibid.). Thus, in these cases, distance does not play a deterrent role.

Sometimes migration of workers determines transport facilities. Migration of large number of workers from Orissa to Gujarat has led the Indian Railway to introduce two direct trains to Gujarat from Orissa.

Both variables—contacts and distance—reflect the psychic costs associated with migration. That is, intervening factors like pecuniary costs involved and the social and psychological costs associated with separation from the native place can deter a move. Friends and relatives in the new location are emotional comforts. On the other hand, long distance between the new (destination) and old location (origin) reduces the chance for frequent visits to the native place. However, both variables also appear to reflect economic moving costs. Destination contacts, as mentioned earlier, provide information and temporary support to the migrants.

As distance increases, direct transportation and migration cost also increases. However, these costs are usually small. It is very small fraction of the higher income required to induce a migrant to move long distances. There is weak support for the notion that distance is also a proxy for information. However, distance and contacts, another proxy for information, are sometimes, correlated (Yap: 1977).

Bright Lights

There may be social forces which affects the decision to migrate from rural areas to towns such as the availability of urban infrastructure and amenities, that is, the attractions of 'bright lights' of the city (Oberai and Singh: 1983; Rempel: 1996). Few studies also seem to support this hypothesis that migrants are attracted to cities in search of better entertainment or 'bright city light', educational opportunities, medical services, cultural and entertainment activities and many other facilities which are not just there in the villages.

Policies of State

The government policies also have impact on movement of people between rural and urban areas (Oberai and Singh: 1983; Rempel: 1996; Stark: 1982a). The most significant of these are: (i) policies fostering a concentrated growth of industrial infrastructure in towns, (ii) import substitution programmes, and (iii) predominantly urban social service investments, which induce migration from rural areas to urban centres. Moreover, leading international development agencies frequently adopt accommodation policies that aspire to improve the lot of migrants, which, in turn, induces rural-to-urban migration.

Risk

It is, sometimes, said that risk avoidance is a major explanatory variable in the decision for rural-to-urban migration (Stark: 1982a). The individual pursuing agricultural production has to bear some risk per period due to low immunity of traditional agriculture, variability in rainfall and weather conditions, plant disease, attacks by pests etc., which affect both grown and stored crops. The low immunity is especially hazardous in the absence of institutional insurance arrangements.

On the other hand, if an individual migrates from rural area to urban centre, he does not face similar periodic risks. However, in the initial period, risks are very high in urban centre since attempt to enter into high-paying sector may fail. Entry into low-paying sectors may be relatively easy. There is high probability of discontinuity of employment in it because it is vulnerable to market fluctuations and sometimes to eviction since it operates on unauthorised occupied land in some cases. But risks associated with urban employment diminish with time and may be relatively lower than the risk associated with agricultural production after some initial, high-risk period. Therefore, it is argued that a rural-to-urban migrant under such circumstances is one who attaches a premium to an early resolution of much of his life-long income-associated risks and he trades in 'medium-level' risks for immediate higher but thereafter lower risks (Ibid.).

Urban/Rural Income

The people migrate for economic gain from poorer areas to richer areas. Difference in income or wage levels between the two places is significant variables affecting movement to locations (Greenwood: 1975; Rempel: 1996; Yap: 1977). It is observed that in some studies that when wage or per capita income differentials are included explicitly, the rate of migration increases with the size of the differential (Yap: 1977).

However, if the notion of wage differences is interpreted normally as differences in money wage, the hypothesis that wage differences are the main determinants of migration has not been confirmed. If the notion of wage differences is interpreted as real wage differentials taking into account the difference in cost of living, various facilities, then the wage difference determining migration has been observed in empirical studies (Greenwood: 1975).

When viewed from the prospective of the rural household, the likelihood of migration of a household member varies directly with the household's perception of the level of income available in urban centre and the perceived probability of being able to obtain such urban income. On the other hand, it varies inversely with the average level of rural income and the equality of access to rural productive resources (Rempel: 1996). Important changes in migration process appear to accompany urban economy decline (Potts: 2000).

Low Agricultural Income and Agricultural Unemployment and Underemployment

Low agricultural income and agricultural unemployment and underemployment in many developing countries are the major factors pushing migrants from rural areas to the areas with greater job opportunities (Oberai and Singh: 1983). The pressure of population, resulting in a high man-land ratio has been widely hypothesised as one of the important causes of poverty and rural out-migration. But, besides population pressures, the low rate of investment in agriculture, fragmentation of land ownership, inequalities in the distribution of land and rural productive assets, allocative mechanism which discriminate in favour of owner of wealth and a pattern of investment and technological change which is biased against labour are other important causes of poverty and rural out-migration.

There is unequal distribution of rural assets in most developing countries. Unequal distribution of land usually encourages large-sized operational farms which have adverse consequences on manpower absorption. Further, it may have adverse effects on the rural income distribution. Thus, more the unequal distribution of land, the greater would be the income and wealth inequalities which would further reduce the carrying capacity of land. The unequal distribution of income arising from an unequal distribution of productive assets is reinforced by the operation of price mechanism due to the prevailing capital and labour market structure (Ibid.).

For these trends, large landowners get richer while the small farmers who find the new technology expensive and more of a risk than traditional method are left behind. So, many small farmers are compelled to sell out their land to large landowners and join the wage labour for work. Therefore, the continuation of the highly unequal ownership of land resulted in increased landlessness and near-landlessness in many developing countries. These changes and consequent changes in social and economic relations induce many of the landless to migrate from their native places either to more prosperous rural areas or to towns and cities (Ibid.).

Migrants in Destinations (Towns)

The studies on determinants of migration indicate that persons move to places with better economic opportunities. Migrants' employment experience on arrival, income gains and economic status relative to urban-born relatives are observed in studies of migrants in individual towns, in groups of towns or in areas within a town/city.

Labour Force Participation

The relationship between migration and labour force participation in towns and cities is indeterminate in theoretical literature (Oberai and Singh: 1983). According to some studies, rural-to-urban migrants are likely to have a lower propensity to participate in the labour force than natives (Ibid.). Several reasons are cited for this. First, migrants may be discriminated against in their search for work because of ethnic, religious, tribal, caste, language, regions and similar consideration. Second, they may be less preferred by the employers since migrants are often considered to have a lower degree of commitment to work than natives. Finally, migrants are not likely to have contacts to the same extent as urban-born natives which may reduce their chance of finding suitable job and thereby place them at a disadvantage in searching for work.

On the other hand, several studies suggest that migrants are likely to participate in labour force to a greater extent than natives (Ibid.). They gave several reasons for this. First, migrants are less likely to have the same degree of support of friends and relatives in towns as they had in their rural native places and are, therefore, under greater pressure to join the urban labour force. Second, migrants, being a marginalised group, often have lower levels of aspirations and expectations and are, therefore, likely to take up whatever jobs are available. Third, migrants except students and retired persons tend to have higher participation rates since many of them are likely to be single persons who have generally higher participation rates. Finally, migrants are an achievement-oriented group and should, therefore, have a higher participation rate.

Marginal Employment

There are also some concerns that rural-to-urban migrants may be limited to marginal employment even though they improve their income by moving (Yap: 1977). However, the scattered evidence does not support this contention. Average income and unemployment rates for migrants and urban natives of comparable age, education and sex are very similar. Aggregate income comparison between migrants and natives shows that the income of urban dwellers is more a function of education or skill levels than of migration status. Underemployment is usually a more serious problem than unemployment for unskilled workers.

When gender aspect is taken into consideration, it is observed that women workers in urban informal sector are engaged mostly in the lower strata of activities, and in terms of occupation, women are largely engaged in personal services. Hence, the scope for upward income mobility is quite limited among many of the women workers (Mitra: 2005). However, long-duration migrants and natives are relatively better off, because they manage to access jobs which yield higher earnings, possibly through their acquaintances and a greater degree of familiarity with urban labour market (Ibid.).

Informal sector is an important source of employment for migrants. Informal sector in urban centres offer migrants a better source of livelihood compared to rural avenues. There is also upward mobility of IS workers, though in a limited sense (Mitra: 2006). However, there is little evidence to show that migrants are disproportionately concentrated in IS (Mitra: 2005). If IS is defined by industry and includes construction, commerce and services, the proportion of migrants is not higher than that of non-migrants in this sector. If it is defined by size of firm, the same conclusions are derived. Nevertheless IS does seem to perform an assimilation function. It usually absorbs new migrants.

Policies

For many, migration has been a livelihood strategy. Therefore, there is a case for genuine policies to support those involved in migratory movement particularly seasonal migration in rural areas, rather than focusing on hampering them (Haan and Rogaly: 2002). Policies that restrict migration are costly and, they hurt the poor more than the rich. Migration should be seen as the norm rather than the rule, as an integral part of societies rather than a sign of rupture—an essential element in peoples' livelihoods, whether rich or poor (Haan: 1999). Policies should aim at enhancing its contribution to people's livelihoods. This includes minimising discrimination against migrants, settlement of immigrants, making remittance easier and facilitating channels to invest these in productive and sustainable production (Ibid.).

It also includes reducing the costs and stresses of migration to migrant workers, via the provision of information about jobs, or providing support to the most vulnerable migrants, and facilities that enable parents to combine employment with child-care and other family responsibilities (Haan and Rogaly: 2002).

Studies on Rural-to-Urban Migration and IS in Orissa

There are some studies on rural-to-urban migration. One deals with macro-economic factors and others depend on micro data. Among the micro studies, few are in destinations and others are in origins and one in both origins and destinations.

Migration and Urban Growth

There are essentially three components of urban growth: (i) natural rate of growth of population, (ii) rural-to-urban migration, and (iii) reclassification of rural areas as urban. While natural growth is considered as an important component of urban growth by demographers, economists feel that rural-urban migration is the major component and that natural growth of population is implicit as a large proportion of the migrants consist of young adults who are in the high fertility zone. Two principal hypotheses are advanced to explain the transition from a traditional rural society to a modern urban society. One, that urbanisation and rapid growth of any city takes place because of an unusually rapid rate of population growth exerting a pressure on limited farm acreage, thereby pushing the landless labour into cities and towns. The second hypothesis emphasises economic forces which pull migrants into urban centres (Williamson: 1988).

While there are a number of hypotheses as to what drives migration and urban growth in developing countries, two principal hypotheses have been advanced: (i) socio-economic transformation 'pushing' rural labour towards cities and towns, and (ii) strong economic forces 'pulling' migrants towards urban centres. The fundamental forces 'pushing' and 'pulling' migrants towards cities and towns in developing countries are classified under three headings (Becker et al.: 1986, 1992; Williamson: 1988, 1991): (i) endogenous limits to city growth such as public social overheads and private housing investment and urban cost; (ii) exogenous external forces such as: (a) foreign capital, (b) relative prices of traded goods in the world market, and (c) fuel prices over which the country has little control; and (iii) exogenous internal events such as: (a) scarcity of arable (agricultural) land, (b) Engel effect, (c) urban bias, (d) government manipulation of the capital market, (e) technological changes and unbalanced growth in factor productivity favouring the urban modern sector, and (f) Malthusian Law (Williamson: 1988).

In the study dealing with macro-economic factors particularly seven sets of variables, viz., (i) price trends, (ii) production conditions, (iii) technological changes, (iv) labour force, (v) consumption pattern, (vi) rural socio-economic transformations, and (vii) natural calamities, responsible for rural-urban migration, it is observed that the slowdown in rural-urban migration and thereby urban growth in Orissa since the late 1960s are mainly attributable to the declining strength of both 'push' and 'pull' factors in the state of Orissa (Samal and Meher: 1997; Samal, Meher and Jena: 1998).

The findings show that the twin macro-economic factors of unbalanced production against manufacturing and rising capital intensity explain much of the slowdown in rural-to-urban migration and thereby urban growth during 1980s, compared to that of 1970s. During 1973-74 to 1986-87, the slow down in rural-to-urban migration and urban growth is mainly explained by increasing capital intensity and the falling rate of labour force (Ibid.). Change in fuel price is not found to be significant determinant of the decline.

Along with weak pull factors, the socio-economic transformation in rural Orissa has been such that the push factors have indeed been weakened. The average size of arable land holding declined at slower rate during the late 1980s than during the early 1970s. During the same periods, inequality in land as well as rural asset distribution has not worsened—the status of rural households has rather improved. The weak push factors may also be due to (i) government sponsored rural development programmes, and (ii) the growth of RNFS in Orissa (Ibid.).

Five Class-I Towns and Five Villages

It is observed from various studies that there is a link between rural-to-urban migration and UIS. While probabilistic models like that of Todaro emphasise pull factors in the process of rural-to-urban migration, the research of political economy gives importance on push factors. Further, rural-urban current earning differential provide a more plausible explanation for migration than rural-urban differences in expected lifetime earnings. There is also little graduation from IS to FS in an urban economy.

Few studies have exclusively dealt with graduation hypothesis in India. That too, these studies have examined only the forward movement of workers from IS to FS in an urban area but no serious study has ever analysed whether workers in urban informal sector (UIS) instead of moving upward to FS, have moved downward to rural informal non-farm sector or rural agricultural sector. Moreover, the society and socio-economic conditions of origins which the rural migrants in UIS quit are not directly surveyed or considered particularly with comparison to that of non-migrants in their native villages to know whether the migrants are the poorest of the poor in their villages. But these are most vital aspects of migration of rural workers to UIS which are mostly neglected by earlier studies on the subject in India.

Considering the above deficiencies in the earlier studies, one of our studies (Samal and Meher: 1997, 2002) sought to answer the various issues from our field data such as whether UIS exists only due to rural-to-urban migration? Whether migration from rural areas to UIS is owing to pull of urban centres or push from rural areas? Are the rural migrants in UIS poorest of the poor in the place of their origins? Do the rural migrant workers in UIS graduate to FS or move backward to rural informal sector or agricultural sector?

To answer these questions and other related issues, we (Ibid.) attempted to study IS with special reference to migration from rural areas to five class-I towns in Orissa, viz., Berhampur, Bhubaneswar, Cuttack, Rourkela and Sambalpur. While Cuttack, Berhampur and Sambalpur can be broadly termed as commercial and administrative towns; Bhubaneswar and Rourkela are designated as modern, planned towns. Of course, Rourkela is an industrial town while Bhubaneswar is the capital city of the state.

Five villages of the sample migrant workers in UIS were selected to study the socio-economic conditions of the migrants in their natives (Ibid.). The villages were selected on the basis of higher frequency of workers present in UIS (more than one worker from the village working in UIS). A part of each selected village was studied which consisted of the cluster of the sample migrants mainly on the basis of caste groups. The villages were: (i) Jakarmarpalli of Ganjam district from where two sample workers were working in UIS of Cuttack, (ii) Aladigam of the same district in the southern part of Orissa (two sample workers in Bhubaneswar's UIS), (iii) Digsira of Bolangir district in the western (two sample workers in UIS of Sambalpur), and (iv) Olagam and Prustisahi of Bhadrak district (three sample workers in Rourkela's UIS, one from the former and two from the latter village).

A mixed sampling was used to study the migration of workers to UIS (Ibid.). In the first stage, UIS was purposively selected in the five class-I towns of Orissa. The quota sampling method was adopted to select migrant workers in UIS of five class-I towns of Orissa. IS enterprise was defined as one, which satisfies at least one of the following characteristics: (i) employ less than 10 workers, (ii) not registered, and (iii) does not maintain complete account. First, the population of UIS units was stratified into two broad strata, viz., establishment sector and non-establishment sector. Each stratum was further subdivided into sub-sectors (first sector into shops, hotels and commercial establishments and only two particular occupations, i.e., rickshaw-pulling and construction work were chosen from the non-establishment sector) and a sample was drawn from each sub-sector on the basis of the given quota to the interviewers. The chosen sample was mostly hired workers in the establishment sector. The size of the quota from the two occupations in the non-establishment sector was made on random basis to limit the sample. The FS units were selected on random basis and thirty sample workers were selected from FS in each town. Mobility of workers from UIS to FS or to rural areas was studied in two ways (i) by enquiring the FS workers about their previous jobs, and (ii) by asking the UIS workers about the non-sample UIS workers who were either their colleagues or neighbours (Ibid.).

The data for the study (Ibid.) was collected from both primary and secondary sources. The primary data was collected by personally administering questionnaire to workers in UIS and FS through specially designed questionnaire. Furthermore, primary data from the households of five villages or part of them was collected through structured schedules.

The major findings of the study (Ibid.) were as follows:

UIS and Rural-to-Urban Migration: First, the study confirmed the close link between UIS and rural-to-urban migrants as more than two-thirds (around 71 per cent) of UIS workers were rural-to-urban migrants. The UIS exists mostly due to migrant workers in case of Rourkela and Bhubaneswar, the two modern, planned towns in Orissa; while in Cuttack, Sambalpur and Berhampur, which are mainly commercial and administrative towns, the share of migrants were less than seventy per cent. Comparatively higher share of migrant workers in case of Rourkela and Bhubaneswar is due to attraction of a giant public sector steel plant in the former and that of expanding capital town in the latter. The pull factors may be powerful for the rural migrants in the early stage of development of the planned, modern towns. Moreover, the local tribal of Rourkela area who were displaced did not take up the informal activities; and in Bhubaneswar, the number of natives was insignificant since the town was constructed and has been expanding on vacant government and forest land. So, most of the inhabitants of these towns are migrants and the two towns have the power to create centripetal force in bringing more migrants. The commercial and administrative towns do not seem to have a strong centripetal force.

Rourkela attracts not only migrants from within the state but also from other states. This is because of the cosmopolitan nature of the town, even though, higher percentage of migrants was inter-state migrants, the migrant from the parent district (i.e., Sundargarh) have a lower share. On the other hand, Samabalpur has a greater role in attracting inter-district migrants. Significantly, lower proportion of intra-district migrants are found in Samabalpur's IS followed by Bhubaneswar and Rourkela. The reason for lower proportion of intra-district migrants in case of Rourkela and Bhubaneswar is understandable. Rourkela is a town nearer to the border of Bihar (now Jharkhand) and migration from Bihar/Jharkhand to this town is significant. Moreover, Sundargarh district in which Rourkela is situated, is dominated by tribals who usually do not like to leave their native places in natural environment to work in UIS. In undivided Samabalpur district, rural hinterland except Padampur subdivision is developed; the southern part of the district due to developed agriculture consequent upon canal irrigation system and northen part for industries. Moreover, due to the presence of Hirakud Dam, leading educational institutions and industrial units in the satellite towns of Burla and Hirakud surrounding Sambalpur have a greater role in attracting inter-district migrants. The same argument may be applicable to the undivided Puri district where Bhubaneswar (now in Khurda District) is located. The very lower share of inter-district migrants in Berhampur's IS is due to absence of large FS units in the town. The migration of workers to UIS (except Rourkela in Orissa) is mostly a phenomenon within the state due to working of centripetal force on the one hand and lack of cosmopolitan nature of the towns on the other (Ibid.).

Determinants: Push vs *Pull*: Second, there are two types of determinants—immediate and secondary— of migration from rural areas. There is little agreement about the immediate determinants of migration within or away from rural areas. There are three different approaches to the materialistic basis of migration. While the neo-classical traditional writers tend to view migration as a kind of 'profit-maximisation' activity, researchers interested in circulation have tended to emphasise 'risk minimisation' and, argue that migration is often a way of diversifying economic options. Researchers who take structural perspective usually interpret migration as a 'household survival strategy'. The concept of 'survival' emerges within this approach as a way of focusing attention on the poverty of many migrants and on the fact that their movements often result from deteriorating employment and income conditions in rural areas (as a result of land concentration, farm mechanisation and low price for peasants' crops). Migration is, therefore, a response to and a strategy for managing this decline in opportunities. The major difference between theoretical models of migration concerns the views on secondary determinants. The dominant model of migration determinants for past three decades is generally referred to as Harriss-Todaro approach.

The major contribution of political economy to migration research has been cumulative evidence it has provided on ways in which international market forces, national, social and political structures and development policies and strategies tend to structure economic opportunity and hence the distribution of labour within developing countries. Evidence of this kind is so overwhelming that Todaro and other scholars in the field who initially ignored these matters have subsequently tried to incorporate elements of political economy into their models and analysis.

Though, there is controversy about the push and pull factors, both the factors have been responsible for massive migration. The 'push' and 'pull' are simply evidence of wider socio-economic condition and structural transformation in the society.

The push factors rather than pull factors, as a prime cause of migration from rural areas were observed in five class-I towns of the state as a whole (Ibid.). While a little less than half of the rural-to-urban migrant (RUM) workers came to the towns being pushed off from their rural native places due to poverty, inadequate income, unemployment and underemployment; less than one-fourth of the RUM workers was pulled towards the towns due to the attraction of higher current money income and year-round employment facilities as well as higher expected life-time income.

While the push factors had greater weight in Bhubaneswar, Cuttack and Sambalpur, the pull factors were more important in Berhampur. On the other hand, neither push nor pull factors had greater weight in Rourkela's informal sector, but 'following primary migrants' of the town played a greater role in the migration of the RUM workers in the town. However, when push and pull factors are compared, the former has a greater strength among the workers' decision to migrate to Rourkela's IS. If we look at the picture of all the class-I towns as a whole, the push factors are important determinants of rural-to-urban migration followed by 'following the primary migrants' and pull factors (Ibid.).

It is higher current money income in the towns compared to rural native places rather than more expected life-time income that has a greater role in migration decision. Though only one-tenth of the RUM workers expressed their motives of migration as higher current money income in the town, a good percentage of migrants had actually higher current money income in the towns at the time of migration from the rural areas to the towns. In contrast, a very few RUM workers had the motive of more expected life-time income in the towns. This was observed to be more or less the same in different towns under our study except Cuttack (Ibid.). The expected life-time income hypothesis does not seem to have much importance in our study due to the fact that it is too difficult on the part of the workers who are not well educated to calculate the life-time expected income in both native places and towns.

Movement and Mobility: Third, a basic hypothesis of probabilistic model is that IS workers graduate into FS after certain period of job search. But the urban labour market structure prevailing in developing countries is not properly described in these models. Some studies (Banerjee: 1983; Breman: 1977; J. Harriss: 1982; Samal: 1990a) in India show that there is little mobility between IS and FS.

The graduation hypothesis of probabilistic migration model (i.e., mobility of IS workers to FS) found some place in our study (Samal and Meher: 1997, 2002), yet due to the absence of adequate number of establishments in FS, the probability of graduation was found to be low. On the other hand, the study found backward movement of some IS workers to rural areas due to higher cost of living, as well as, the weak probability of getting a FS job in the town. They join either agricultural sector or non-farm activity in rural areas after their backward movement. More than one-third of the RUM workers had moved from one unit to other within IS and around one-fourth of the sample FS workers had graduated upward from the town's IS to their present FS jobs. There was upward mobility of some non-sample workers from UIS to FS in all the five class-I towns (Ibid.).

RUM Workers in Native Places and Towns: It is sometimes argued that only the poorest of the poor in the rural areas migrate to urban centres. This argument may seem plausible due to the significant role of the push factors in deciding rural-urban migration. But, in reality, this is, always, not the case. Our study (Ibid.) found that around half of the households were landless labourers and the rest had some land and/or petty business. But the proportion of landless labourers migrating to urban centres was significantly lower than their share in total households.

This indicates that it is not only the worse-off but also better-off households (in the villages) from which the people migrate to urban areas in search of jobs. Furthermore, the better-off households are having a significant share in the migration stream. This may be due to the fact that (i) they have initial required fund to bear the transport cost, as well as, the maintenance cost for few days of waiting period in towns; and (ii) these households have the capacity to bear loss of income foregone due to the absence of the migrant members (Ibid.).

The non-establishment sector—the lower segment of UIS—was crowded with workers from the lower strata of the caste society, while the establishment sector—the upper segment of UIS—was dominated by the upper strata/forward castes of the society (Ibid.). Thus, the inequality in society particularly in rural areas outflows to UIS in urban centres of Orissa where the workers belonging to SC/ST do not have easy access to the upper segment of UIS.

The income of the own-account workers in UIS in Orissa was significantly higher than the wage of hired workers (Ibid.). Within the category of hired workers, the income of non-establishment workers was higher than that of the establishment workers. Therefore, the non-establishment workers most of whom were illiterate and belonged to the lower strata of the rural society earned income higher than that of the establishment workers who were mostly educated and of forward caste origin. So, in UIS, neither education nor caste factor played an important role in determination of wage/income, though it determined the status of the UIS job, which the UIS participants were able to join. Hence, though social inequality of rural areas continued to surface in UIS in the form of division of labour, there was a reduction of income inequality due to migration to urban centres (Ibid.).

Thus, the contention that only the poorest of the poor in rural areas migrate to towns and cities is not necessarily true.

It is not only the worse-off but also the better-off households in the villages migrate to urban centres in search of jobs. Moreover, the better-off households are having a significant share in the migration stream. Social inequality of rural areas continues to surface in UIS in the form of division of labour among the RUM workers. However, there seems to be a reduction of income inequality among them due to the migration to urban areas. Thus, rural-to-urban migration as a tool to alleviate rural poverty may not easily be challenged (Ibid.).

Distance and Migration in two Modern, Planned Towns

An additional research work (Samal and Meher: 1998) to our original research project (Samal and Meher: 1997) analysed the role played by geographical distance between origin and destination in rural-to-urban migration and migrants' link with their native places covering two modern, planned class-I towns, viz., Rourkela and Bhubaneswar, out of our five sample class-I towns of Orissa. Rourkela is an industrial city in the northern extremity of Orissa and, Bhubaneswar is the capital of Orissa and is situated in the eastern coastal belt. The major findings of our study (Samal and Meher: 1998) are as follows:

Almost all (96 per cent) the sample RUM workers in the two towns as a whole visited to their rural native places on an average of eight times per annum. The expectation of higher frequency of visit with small duration of stay per visit was realised in case of the RUM workers whose native places were at smaller distance from the towns. With the increase in geographical distance of the rural native places from the towns, the frequency of visits declined and the duration of stay per visit increased in case of our sample UIS workers in two towns as a whole. This was also observed in case of each individual town. The main purpose of visit of workers to their native places was just to visit the relatives and to attend the social ceremonies and festivals (Ibid.).

Family ties and inheritance are the key determinants of the remittance of migrants in informal sector (Duraisamy and

Narasimhan: 2000). Remittances play an important role for many areas of origin and, have huge potentials for development particularly since much migration remains circular and migrant households' strategies remain based in villages of origin. However, negative effects of remittance cannot be ignored (Haan: 1999).

More than three-fourths of the RUM workers of two towns as a whole remitted certain percentage of their earnings to their native places (Samal and Meher: 1998). However, the proportion of migrants, remitting money to their native places, declined, with increase in duration of stay in the towns. This indicates that once a migrant has lived away from his rural habitat for a long time, the native bond in respect of financial linkages necessarily weakens; perhaps, he might be bringing his family members in phases to the town. But, as mentioned earlier, the RUM workers were keeping social link with their native places. The percentage of the RUM workers remitting money was higher in the sub-sectors of hotel, rickshaw-pulling and construction as compared to that in shops and commercial establishments. This is plausible, as most of the workers in the former category were socially and economically backward, and their families in rural areas were depending on their remittance. On the other hand, the workers in the latter category were educationally and economically well-up for which their family members in the rural areas were not so dependent on them. Moreover, they felt that they were under-employed and under-paid which did not permit them to remit money to their native places (Ibid.).

It was also observed that more than one-third of the RUM workers of two towns as a whole were bringing cash or kind from their native places. The average amount of cash/kind brought is, however, very insignificant (Ibid.).

Around two-thirds of the RUM workers in the two towns as a whole had come from a distance of more than 100 kms. The long-distant (above 100 kms) migrants in IS of Rourkela were mostly from the coastal belt of Orissa. Hence, there is a close relationship between the number of migrants in IS and the distance of the native places from the towns (Ibid.).

Migrants in Destination (A Class-I Town)

In a micro study in destination, that is, Sambalpur, a class-I town, it was observed that most workers in IS of the town were migrants (Samal: 1990a, 1990c). The study wanted to test by examining the empirical data, the following hypotheses: (i) informal sector is merely a residual mainly absorbing the excess supply of labour in urban area due to migration from rural areas; (ii) migration is primarily on rational economic consideration and the decision to migrate depends on 'expected' urban-rural real income differentials than rural-urban current earnings differentials; and (iii) urban informal sector acts as absorber of migrants and as vestibule for them to enter formal sector.

The geographical coverage of this study (Ibid.) was limited to Sambalpur which is a class-I town in the western part of Orissa with an effective sample of 180 workers consisting of 108 own-account workers, 72 wage/hired workers and the unpaid family labourers, one each from 180 IS units spread over the six sub-sectors of IS of Sambalpur town, viz., manufacturing, trade, transport, service, allied and construction (Samal: 1990c).

The study (Ibid.) shows that about two-thirds of the IS workers as well as of the IS participants were migrants. The proportion of migrants was comparatively more in hawking and hotel trade sector and also in the transport and construction sector. Moreover, if the second and third generation migrants, returned migrants and commuters were included, then the migrants as a whole constituted around three-fourths of the total workers. Thus, a strong link between UIS and migration was found. Equally strong link was also found between migration and FS in the town.

There is a distinction between 'motivation' and 'cause' of migration. In reality, the individual motives and functions of a man are determined by the socio-economic system surrounding him. An individual might have a motive to migrate but that motive is always shaped by social, economic and structural transformation in the area in which he resides, which can be regarded as the cause of migration. Inadequate income and poverty in the native place mentioned by forty-four per cent of migrants and constituting about one-third of the responded total frequencies of multiple responses were found to be the most important causes for migration out of seven different factors suggested in the questionnaire in our study (Ibid.). When we combine different factors responsible for migration under two heads 'cause' and 'motive'—, a comparatively higher percentage of responses and of migrants moving out was due to former factor of migration. Again, when we group the same seven alternatives under three heads—'push', 'pull', and 'others', similarly a higher proportion of responses and of migrants moving out was owing to 'push' factor.

In general, various factors show that the socio-economic conditions of rural labourers has been deteriorating gradually over a period of time which may indicate a 'push' as the 'cause' of migration from rural areas. Most of the migrants were from rural areas. Around half of the migrants were inter-district migrants. Among the inter-district migrants, around three-fourths had come from neighbouring districts. It is to be noted that around four-fifth of inter-district migrants from neighbouring districts moved out of the drought-prone districts of Bolangir and Kalahandi (Ibid.). But, only one worker in our sample migrated from the neighbouring district of Sundargarh which is most industrialised in the state of Orissa.

Thus, like the cause and motive of migrants showing the importance of push factors, the analysis of wider socio-economic conditions of rural areas in general and neighbouring districts in particular shows an apparent evidence of 'push' factor as a 'cause' of migrants to move out of their villages (ibid.). In fact, it is the socio-economic conditions of society which is responsible for increased mobility and movement. The 'push' and 'pull' are simply evidence of wider socio-economic conditions and structural transformation in the society. Of course, it may be difficult to bring a distinction between 'push' and 'pull' since both imply that distinction is seen as having an economic advantage over the origin. As is also revealed by our study (Ibid.), the presence of pull factors cannot be denied. Yet, the 'push' is found to be stronger than 'pull'. This is particularly relevant in case a migrant is working in IS. Given low earning jobs and uncertainty and insecurity involved in the work in IS, migration to work in this sector has to be mainly due to unfavourable conditions at the place of origin pushing the worker out rather than attraction of working and living at the place of destination.

Further, four-fifths of the migrants had link with native places through visits, about one-third through remittance, and more than half possess land, among them more than two-thirds have their own cultivation arrangements and virtually none disposed of land even after a period of stay at Sambalpur for an average period of about 13 years (Ibid.). These facts suggest that the migrants in IS had not permanently settled their mind in their jobs in the town. Perhaps, they thought that if their income and employment situation would worsen they might go back to their villages. So, it seems that it is not the permanent life-time expected income which attract them to the town, but their worsening economic conditions in the native place which push them out, though there may be immediate gain in the urban areas of destination due to migration. Most of the migrant workers had immediate gain just after migration and the current urban-rural income difference is more important than the life-time expected income. This study (Ibid.) castes doubt on the life-time expected earning hypothesis to the extent it implies that most of the migrant workers may work in a job which brings them an earning even lower than those in place of their origins since they are aware of the fact that their earnings during the whole working life in the destinations will be much higher than that in their native rural places.

The mobility of workers is understood as purposeful movement of workers from one job to another in pursuit of certain well defined objectives. There was little vertical mobility of workers from IS to FS. The potential mobility from IS to FS was also very low. As a higher proportion of migrant workers who entered IS in the town were fully or partly satisfied and since very negligible number had made some effort to change their present job, it can be safely concluded that the migrants do not consider their employment as a means of survival while waiting and searching jobs in FS. Thus, there was little mobility—both actual and potential—of workers from IS to FS (Ibid.).

Thus, most workers in IS of Sambalpur were migrants. But, it does not necessarily mean that predominance of migrants among workers is a distinguished characteristic of IS, as the proportion of migrants was found no smaller in the total urban workforce of FS. The push factors in rural areas, rather than pull factors, constitute the main factor in the process of rural-to-urban migration. Yet, the general proposition that work in IS is the desperate choice to which they are driven and, therefore, IS is a 'residual' sector does not seem to be valid. In fact, work in IS does provide a certain degree of 'pull' to the migrants. Employment and earning opportunities in this sector provide a positive 'current urban-rural income differential', and migration has, therefore, not necessarily be motivated by 'expected life-time urban-rural real income differential'. That the life-time expected earning hypothesis is of little relevance, also demonstrated by our finding that there is little actual or potential mobility of migrant workers from IS to FS. The graduation hypothesis which assigns to IS the role of vestibule for rural migrants to enter FS, as the ultimate destination, also, therefore, seems doubtful (Ibid.).

Migrant IS Entrepreneurs in Emerging Industrial Centres

In one of our studies (Samal: 1996, 1998) portraying the profile of IS around NALCO (both NALCO Nagar in Angul and Damanjodi sector); we found that around half of the sample IS entrepreneurs in NALCO Nagar and over eighti-five per cent in Damanjodi sector were migrants, most of them coming from the neighbouring districts.

Prior to migration, none of the entrepreneurs in Angul sector was involved in cultivation or traditional non-farm activities. Around half of them were engaged in modern non-farm activities, around one-fifth were agriculture and non-farm labour and the rest were unemployed. In Damanjodi sector, more than half of the IS entrepreneurs were previously in modern rural non-farm activities, one-fifth unemployed and the rest agriculture and non-farm labour (Ibid.).

Majority of the 'migrant' entrepreneurs in Damanjodi sector migrated to start an IS unit around NALCO. One-third of them decided to migrate due to poverty and low level of income. However, migration to Angul sector was not due to poverty and inadequate income at the place of origin of the migrants. They were rather 'pulled' towards NALCO Nagar's IS by opportunities created by the establishment of NALCO's Angul sector in the early 1980s. Hence, it is natural that the IS entrepreneurs in NALCO Nagar had come with the required capital to start business.

As far as Damanjodi sector is concerned, the benefits of the growth of IS activities accrued mainly to the migrants, particularly from the neighbouring undivided district of Ganjam, and not to the local people, who are mostly tribal. The locals could not gain from these activities as they lack both initial capital for investment and entrepreneurial skills (Ibid.).

The study (Ibid.) reveals a significant phenomenon with respect to migration from rural areas to emerging industrial centres which are not necessarily urban centres. Earlier studies in Orissa (Samal: 1990a, 1990c) argued that rural-urban migration of workers to urban IS was mostly due to 'push' factors. The observations from the two sectors suggest that the 'pull' factors are equally important in explaining migration from rural areas to upcoming industrial centres of entrepreneurial individuals (Samal: 1996, 1998). Interestingly, almost all the IS entrepreneurs in Angul sector were pulled towards NALCO Nagar's IS by the opportunities opened up by the Company. The same people might not have migrated to other 'old' urban centres of the state as the scope of starting an IS unit is limited and competition from both FS and the existing IS units is fierce (Ibid.).

Migrant Workers in a Coal Mine Region

Another study (Samal and Mishra: 1998a, 1998b) dealt with the determinants of migration of workers to the coal mine region of Orissa, particularly Talcher coal mine region and the distribution of income among migrant workers in informal sector. An informal sector enterprise is one which satisfies at least one of the following characteristics, according to ILO (Hussmanns: 1995): (i) it employs less than 10 workers, (ii) not registered, or (iii) it does not maintain complete account.

The workers engaged in mining did not come under the purview of the study (Samal and Mishra: 1998a, 1998b). In particular, the study identified three categories of workers: (i) informal workers in formal (coal mine) sector (IWFS), (ii) informal workers in informal sector (IWIS), and (iii) entrepreneurs in informal sector (ENIS). IWFS refers to those workers who are involved in activities (of FS establishment) which are given to private parties who invoke informal workers to carry out the activity. The development of any industrial activity/mining activity (in this case, coal mining) would create a category of people (say, salaried employees in FS) who generate a demand for different goods and services, which in turn, gives scope for creation of large number of activities in IS, thus opening up job opportunities in the same. The workers (employees) in this sector are referred to as IWIS whereas the owners/employers are referred to as ENIS (Ibid.).

The main objectives of the study (Ibid.) were (i) to examine the distribution of income among migrant workers in IS, and (ii) to analyse the factors that may influence the migration of IWFS, IWIS and ENIS. Keeping in mind the above objectives and after a preliminary visit to Talcher, a questionnaire was prepared and, canvassed among 78 respondents—26 IWFS, 25 IWIS and 27 ENIS. Most of the questions were close ended to make the computation and tabulation easier.

The major findings of the study were as follows: Most of the respondents were male and Hindu with Oriya as their mother tongue. The OBC constituted the largest group. Though, most of the respondents were married and had done some years of schooling, a large proportion of workers in informal sector (IWIS) were unmarried and, there was hardly any who has continued his study beyond normal schooling (Ibid.). There were no respondents who could be considered as children (below 14 years) or old (above 58 years).

(a) Distribution of Income among Migrants: As per the weekly money income, the IWES were getting less income than

their counterparts, viz., IWES and ENIS. The main factors which determined the distribution of income among migrants were: (i) distance from destination, and (ii) age of migrant (Ibid.).

First, distance had a positive association with income. Higher the distance between the origin and destination of the migrants, the higher is the income at the latter. The migrants from distant places were able to meet the transport cost and moving cost (in initial transport cost and cost of visits to native places) since their incomes were comparatively higher than short-distance migrants. Therefore, if the income at the point of destination is comparatively high, the transportation cost and moving cost may not deter the migrants to move from their native places. The private benefits of migration were substantial enough to offset the most reasonable direct transportation expenses. However, the rate of return on investment in migration differed appreciably for various sub-groups of migrants, such as workers, entrepreneurs, etc.

Income after migration was more than the income of respondents before migrating to Talcher, as it is hypothesised that the people migrate for economic gain. The difference in income between origin and destination of migrants is a significant variable affecting the migration (Ibid.).

Second, the age-wise income distribution showed that distribution became more and more skewed towards the older people who belonged to higher income groups (Ibid.). That is, workers in the higher age group had a greater probability of earning a higher income. This was particularly the case among ENIS and to a large extent among IWFS since, these two sub-sectors represented the upper strata of the town's IS and, of these, IWFS might be earning more due to the greater work experience and heterogenous nature of these groups. However, this was not very evident among IWIS, because of the absence of any scope for earning higher income among the respondents of IWIS.

(b) *Determinants of Migration*: The motives of migration were classified as 'push' (factors which emphasise on the situation at the origin, that is, place from which migration started), 'pull' (factors which emphasise on the situation at the destination, that is, Talcher for this study) and, 'push N pull' (those which could be referred to as a 'push' as well as 'pull' factor) (Ibid.).The respondents identified about 11 factors while giving reasons for their migration to Talcher. Of these six were 'push' factors (inadequate income in native place, inability to carry out cultivation at native place, natural calamities, disturbing development in the family like death of a family member or quarrel, employer transferred and/or war), three were 'pull' factors (better employment/income earning opportunities at Talcher, relative working nearby, an urge to see new place) and, the remaining two were 'push N pull' factors (employment around the year at place of destination against seasonal employment at place of origin, and the facility to learn a trade which will be useful at the native place). Further, the respondents also ranked their preferences whenever they gave multiple motives. This latter aspect is a novel feature of this study (Ibid.), which helped in making the analysis on migration more rigorous and robust.

There was a clear predominance of 'pull' factors in rank one and rank two. Further, a majority of those who placed a 'push' factor in the first rank had given a 'pull' factor in rank two. The 'pull' factor was more strong in all the three ranks. To make the analysis more interesting, the relationship of motivation with age and educational qualification of the migrants was taken up since, a number of personal characteristics are likely to have an impact on the individuals' decision to migrate.

First, those who had given a 'push' factor as the prime cause for coming to Talcher were relatively older than those who gave a 'pull' factor as the prime (rank one) cause. This can be interpreted in two ways. One is that older people consider moving out from their places of origin under severe adverse conditions—a more general interpretation. The probability that a worker will migrate is likely to decrease as his age increases since older persons have a shorter expected working life over which to realise the advantage of migrating, which makes the rate of return on migration lower for them. Individuals for whom migration is profitable immediately, move without delay because, postponing movement would result in a loss during early years. Moreover, job security and family ties are also likely to be more important for the former than for younger ones which discourage the former from migrating. Thus, older people do not like to migrate to towns even for augmenting their income, so that towns/cities do not 'pull' them. If they, at all, want to leave their native places, it is because they are forced or 'pushed' out of their origins due to adverse socio-economic conditions, natural disaster or other adverse factors. On the other hand, the younger people are more likely to be attracted towards towns/cities for 'bright lights' and because they offer a better working life. The other possible reason is that older people are the ones who migrated long ago thus indicating that at an earlier date probably Talcher did not have much potential for pulling migrants as today.

Second, as far as educational qualifications are concerned, it was illiterate and those who could read and write without any formal education, who had a higher probability of citing 'push' as a prime cause (Ibid.). This could be so because those with relatively less educational qualifications may have low levels of skill and, hence, access only to low income activities and, thus the probability of their getting high-paying jobs in it urban centres is very low. As a result, they are usually not attracted to urban centres. Understandably if, at all, they came, it is due to 'push' factors and almost all of them joined IS. On the other hand, the educated are pulled towards the urban centres, be it IS or FS, because they have better access to information on employment and job opportunities and tend to respond to the same in a positive manner. Education reduces the importance of tradition and family ties and increases the individual's awareness of other localities, which as a consequence, weakens the forces that hold him to his native places. Moreover, the opportunities for getting jobs for the better educated are better than that for poorly educated. In effect, while the better educated are 'pulled' towards towns and industrial centres, most of the poorly educated are 'pushed' first from their native places and then 'pulled' towards the destination.

Thus, 'pull' factors as against 'push' were more predominant in inducing migration to a coal mine region. This finding is robust because the respondents were given the opportunity to identify multiple causes and then rank their preferences. The 'pull' factors are further reinforced by the fact that most of them have their jobs prearranged, may be by their friends, relatives or contractors. In case of entrepreneurs in IS, they moved out of their origin not due to 'push' factors but to earn money by starting businesses as entrepreneurs. They were better-off and had entrepreneurial skill and hence were attracted to Talcher. Despite the predominance of 'pull' factors, one did observe some subtle differences among three occupational categories.

Inter-State Migrants in Origins

There are two studies dealing with inter-state migrants in their origins in the undivided districts of Ganjam and Puri (Mishra: 1998; Nayak: 1993).The studies show that most of the migrants from these two districts were young, illiterate and, from marginal farmer households. The labourers migrate in groups to urban centres in other states mostly with kith and kin or with people belonging to a particular locality. While Mishra's (1998) study shows that the migrants move out of state of their own accord with the help of friends, relatives and other co-villagers; Nayak's (1993) study found that migration of inter-state migrants is controlled by labour contractors and agents.

Summary

Each migration pattern will have consequences that are unique to itself and to the economic context which it brings about. The most popular model is that of Todaro which regards the migration as a two-stage phenomenon by recognising a dichotomy in an urban economy. By incorporating IS into standard Todaro type migration model, Stark postulates a two-pronged planning horizon that rural-to-urban migrants may consider. Various empirical studies show that migration does occur from low to high income regions but some debate continues regarding the relative importance of the characteristics of origin and destination in explaining migration. Various independent variables are used to explain as determinants of rural-to-urban migration such as level of education, personal contacts, geographical distance, bright lights, government policies, risk avoidance, urban-rural income difference, agricultural unemployment/underemployment and others. The relationship between migration and labour force participation in towns and cities is indeterminate, as found in both theoretical and empirical literature. The scattered empirical studies do not support the contention that rural-to-urban migration may be limited to marginal employment even though they improve their income by moving. The studies on rural-to-urban migration in Orissa found push factors, in general, as more important determinant than pull factors in the process of rural-to-urban migration and confirmed the close link between UIS and rural-urban migration. However, pull factors are predominant in emerging industrial centres for IS entrepreneurs and in coal mine region for most of the IS workers. The contention that only the poorest of the poor in rural areas migrate to towns and cities is not necessarily true. However, rural-to-urban migration as one of the tools to alleviate rural poverty may not easily be challenged.

Note

1. "Expected" gains are measured by (i) the difference in real incomes between rural and urban job opportunities, and (ii) the probability of a new migrant obtaining an urban job. Todaro has cited an arithmetical example to explain this.
   Suppose the annual average real income is 50 units per year in the country-side against 100 units in the city for a rural unskilled or semiskilled worker. The probability of securing a high-paying job in urban area within one area is one chance to five. Thus actual probability of success of getting urban job is 20 per cent so that his expected income is 20 units in the one-year period and not 100 units. So, it is not rational for migrant to seek the urban jobs. But if the probability is 60 per cent, the expected urban income is 60 units, and it is rational for migrant to try to get the urban job.
   Todaro has also explained this with a mathematical formulation for life time expected income over a long period time horizon. In this mathematical formula of the model (Todaro: 1969); the decision to migrate or not will depend on whether:

$$V(0) = \int_{t=0}^{n} [P(t)\, Yu(t) - Y r(t) = \ddot{y}\, e^{-it}\, dt - C(0)$$

is positive or negative.

V (0) = the discounted present value of the expected 'net' urban-rural income stream over the migrants time horizon (the bracketed term zero refers to time period zero).

Yu (t) = average real income of individuals employed in the urban economy in time period't'.

Yr (t) = average real income of individuals employed in the rural economy in the time period't'.

n= the number of time periods in the migrant's planning horizon.

(-) = the minus symbol used for future income less present value.

i = the discount rate reflecting the migrant's degree of time preference

t = 0 to ....n, the time horizon.

P (t) = is the probability that a migrant will have secured an urban job at the average income level in period't'.

[P (t) Yu (t) – Y r (t)] = the expected real income differential in time 't'.

[P (t) Yu (t) – Y r (t)] e^{-it} = the present value of expected real income differential in 'tth' year after discount at the rate of 't'

where:

e = base of natural logarithm

= 2.178281828....

$$\int_{t=0}^{n} [P(t)\, Yu(t) - Y r(t) = \ddot{y}\, e^{-it}\, dt$$ = the present value of expected earnings differentials for a migrant in time period 't'. (We get by adding income differentials for all the time periods).

C (0) = cost of migration of the individual migrant

$$\int_{t=0}^{n} [P(t)\, Yu(t) - Y r(t) = \ddot{y}\, e^{-it}\, dt - C(0)$$ = net expected real income differentials; (expected income differentials) - (cost of migration)

If it is +ve, migration is desirable; i.e the individual will decide to migrate.

If it is –ve, migration is not desirable; i.e. the individual will not decide to migrate.

5

Dynamics and Growth

The size and composition of informal sector has varied among urban areas depending on their dominant activity and the rate of growth. Despite these variations, it has been observed that the size and extent of IS tends to increase in most of the urban centres in developing countries (Papola: 1987). Informal sector growth could assume the following forms:

(i) Growth due to rise in the local and migrant labour force in spite of the slow growth of FS: This pattern of IS growth is characterised by activities which have direct linkages with FS and low productivity and low earning levels.

(ii) Growth of IS accompanied by a relatively faster growth of FS: In this case, IS gets increasingly and directly linked with FS through technology and market and, generates employment with a reasonably high level of productivity and earnings, which in comparison with FS are still lower. This line of IS growth is observed mainly in cities and towns experiencing a rapid expansion of industrial activity in FS (Ibid.).

(iii) The growth in units and employment are guided by the survival strategy and macro-economic circumstances. The pattern of increase in employment does not correspond with improvement in performance or efficiency. The credit is the most important factor in the effective functioning of the units (Kundu, Lalitha and Arora: 2001).

(iv) A sudden emergence and growth of IS activities when a giant FS industrial unit is established in a non-urban area due to growth of demand for certain kinds of goods and services emanating from the incomes of the employees of the FS establishment or others related to it either before or after its establishment (Samal: 1996, 1998). The establishment of a big FS unit and growth of IS surrounding it leads to urbanisation of the area. But this IS, in most of the cases, does not get linked through the technology and market (output/input) with FS.

According to one study (Bhattacharya: 1998), in India, during 1970s, IS was a dynamic and productive sector, well capable of attracting and sustaining labour in its own rights. The IS was far from being a mere passive absorber of labour. It played an important role in rural-to-urban migration during 1970s. Although, the influx from rural areas of the state have decreased in Calcutta over the decades, there is a rise in informal activities in Calcutta during the last few decades since migration from other states has increased (I. Mukhopadhyay: 1998). The nature and pattern of labour use in the city have been a process of reallocation of labour force in favour of the self-employed labour force (Ibid.).

It appears that the hypothesis that the level of IS activities are demand-determined (Richardson: 1984) is applicable in case of growth of IS activities around an emerging industrial centre. If the demand for IS goods and services is a linear and homogeneous function of urban income, then:

$$Y_i = aY_u$$

where Y_i = informal sector income
Y_u = total urban income
a = constant

It is widely accepted that the prospects for growth in IS depend heavily on what happens in FS, particularly in industrial centres and industrialised towns and cities. While the service-component of IS is demand-determined, the good-producing component has both demand (intermediate goods for FS firms and consumer goods for FS employees) and supply sides (raw materials and intermediate inputs from FS firms and the provision of working capital through sub-contracting relationship) (Ibid.).

In Orissa, in Angul and Damanjodi sectors of NALCO, the very origin of IS has been due to the demand emanating from company's employees, contractors and others related to it (e.g., the construction workers during the installation stage). In one of our studies (Samal: 1996, 1997c, 1998), which was an one-time survey, we have dealt with the growth of IS around NALCO's two sectors in Orissa.

The integrated National Aluminium Company has five main segments—bauxite mine, aluminium refinery, aluminium smelter, captive power plant and port facilities. The refinery is located at Damanjodi and, the smelter and captive power plants are at Angul, a district headquarters. A small township called NALCO Nagar started growing around Angul sector, 5 kms away from Angul town. Damanjodi sector is in Koraput district, 35 kms away from the district headquarters. This sector is close to the National Highway and spread over 53 revenue villages. The native population is predominantly tribal.

There has been rapid growth of IS activities, mainly shops and commercial establishments, around two sectors especially so in NALCO Nagar. The number of shops and commercial establishments in Angul sector at the time of survey for the study in 1995 was 277 and 79 respectively. In Damanjodi, there were 154 shops and 48 commercial establishments in IS.

While some IS units in NALCO Nagar was selling directly to NALCO, not a single unit in the sample in Damanjodi sector was able to do so. However, most of the total sales turnover of IS of Damanjodi sector and NALCO Nagar was due to demand emanating from NALCO employees and others related to the company. Thus, though the direct backward and forward linkages of IS of both the sectors, with NALCO were negligible, the indirect linkages appeared to be significant. Hence, we conclude that the growth of IS in both the sectors is demand-determined.

The most common cause for closure of informal units, as revealed by the study (Ibid.) is failure of business either due to a fall in overall demand for IS goods and services, or the inability of the units to face competition from other similar units. During the construction stage of NALCO Angul sector, the trade sub-sector of IS (including hotels and eating houses) was growing in an impressive space. They were catering to the demand of the manual workers whose number far exceeded that of the employees of NALCO. After the completion of the project, there was not only a fall in general demand by manual workers, but also a shift in demand from wage goods to different kinds of goods and services required by the workers and white-collar employees of NALCO Angul sector.

In this transition from the construction to the production stage of NALCO, some of the IS units declined while others flourished. Also, some new IS units emerged to cater to the new kinds of demands. Many IS entrepreneurs might have started their units without assessing the exact demand potential. Hence, the supply of some particular goods and services might have exceeded the demand for them, resulting in intense competition and elimination of inefficient IS units (Ibid.).

The establishment of large industrial units thus generates enough opportunities for enterprising people to start self-employment ventures in informal sector to benefit from the newly emerging demand. The nature and pattern of this demand could be different in different stages of the project leading to closure of some units and the survival of others in the face of severe competition.

The benefits of growth of IS are largely distributed locally when the enterprise are owned by natives as was found in Angul sector in our study (Ibid.). When the entrepreneurs are predominantly migrants and the 'locally displaced population' marginalised, the growth of IS does not result in any perceptible benefit to the local people. These are important aspects to be considered by policies aimed at developing areas with large units in FS (Samal: 1997c).

Thus, we observe that there is a sudden emergence and growth of IS activities, when a giant formal sector industrial unit is established in a non-urban area, which are demand-determined.

In another study (Samal; 1994a, 1997b) we have tried to redefine IS by studying the dynamics of informal manufacturing sector (IMS) at two points of time in a class-I town in the western part of Orissa.

There has been a rapid diffusion of the concept of the informal sector since 1971 and more characteristics have been added to it by later studies to get rid of the element of vagueness.

Attempts have been made to define different categories within informal sector. Breman (1976) has classified the urban working population into four broad social classes: (a) employers, (b) self-employed (own-account workers), (c) unpaid family labour, and (d) wage workers. But this traditional categorisation does not take into account various differences among workers. Several systematic attempts have been made to establish a typology of labour status. For instance, Hariss, Kannan and Rodgers (1990) have developed a labour status approach based on the differences in labour status in terms of 'protection', 'regularity' and 'autonomy'. The main finding of their study is that labour market segmentation is partly embedded in the overall pattern of social organisation which determines, along with the dominant labour institutions, the nature of access to the labour market and the distribution of the benefits from economic development. This approach questions the very idea of informal-formal dichotomy on the ground that it is too simplistic a demarcation, and policy formulation and implications based on it are likely to be ineffective. It provides various characteristics of different groups of workers and their mobility, which will be used in this paper to define different kinds of 'establishments' in informal manufacturing sector (IMS).

Importance is given in this study (Samal: 1994a, 1997b) to the composition of labour within an establishment, rather than labour households or the labour market. The emphasis is on the establishments of IMS in a town to refine the concept of informal sector and to avoid confusion arising out of informal-formal dichotomy, particularly in the manufacturing sector.

The approach may be called the labour composition establishment approach. The establishments are classified on the basis of their composition of labour, viz., own-account workers, family labour (both paid and unpaid) and hired workers. The study is based on surveys done at two points of time. The initial benchmark survey was done in 1985 and a resurvey of the same sample units was done in the middle of 1994. The thrust is on reclassifying the sample units for theoretical refinement of the concept of IS after observing their working and operation at two points of time.

The Benchmark Survey

The benchmark study in Sambalpur (Orissa) during 1984-85 concentrated on the economic aspects of all the activities in urban informal sector (Samal: 1986, 1990a, 1990b, 1990d). For the study, selected enterprises were approached personally and data was collected through personal interviews. The size criterion was adopted to demarcate the IS units from the FS ones. Following this, the IS units were defined as those employing ten workers or less, including the owner-operator, paid and unpaid family workers and hired workers. The second-order condition was the presence of certain specific features derived from the size criterion, viz., small-scale operation, and direct participation of the owner of the unit either in manual or entrepreneurial work. Other features are: (a) the unit in general has an informal relationship with the employees and customers; (b) institutional factors have no role in determining wage, which is decided on the basis of bargaining between the employer and employees; and (c) there is no security or certainty in employment and income. However, this study only partially explained the concept of IS as distinct from FS.

The benchmark study included production firms such as traditional/craft-based/artisan units, modern manufacturing and processing establishments, and workshops and garages (including repair and fabrication units). Out of an estimated total of 630 units in the town, the effective sample of establishments in IMS was fifty-eight, which were distributed among ten groups when classified on the basis of the raw materials used, such as earth/cement-based, wood/forest produce-based, foodgrains-based, cloth-based, tobacco-based (*biri*), paper-based, metal-based, and paints and chemicals.

The Follow-up Study

Though the same concept of IMS used in the benchmark survey was adopted for the follow-up study in 1994 (Samal: 1994a, 1997b), three new concepts were introduced to bring some distinction between different groups of informal manufacturing establishments. Following the labour composition establishment approach, the study classified IMS into three distinct categories on the basis of the nature of participants. They are:

(i) *Micro business*: Micro business is defined as a unit/ establishment which is managed by only one person who is the owner/entrepreneur.

(ii) *Petty producer unit*: A petty producer unit is one where the owner/entrepreneur works along with paid/unpaid family labourers in the establishments, the total number of participants being ten or less. There are no wage labourer.

(iii) *Small capitalist producer firm*: A small capitalist producer firm is a unit where the owner employs wage workers in small numbers. The total employees, including the owner-operator, paid/unpaid family labour and wage workers, are ten or less. The apprentice is treated a wage worker.

The manufacturing units surveyed in 1985 were also classified under these categories in the follow-up study in 1994. The follow-up study had the following objectives:

(i) To assess the mobility of units from informal to formal manufacturing sector as also among micro businesses, petty producer units and small capitalist producer units.

(ii) To measure the changes experienced by the units over the nine-year period between 1984-85 and 1993-94, especially in fixed capital, capital-output ratio and gross revenue.

(iii) To analyse the growth of IMS in terms of employment, output turnover and the number of establishments.

(iv) To understand the problems and prospects of these units and to see how far government policy measures can address these problems.

(v) To verify whether any noticeable difference can be found in the nature of the IMS establishments when they are grouped under the three categories of micro business, petty production unit and small capitalist producer unit.

Location of the Study/Data Collection

The geographical coverage was limited to Sambalpur town in the western part of Orissa. Sambalpur is a district and divisional headquarter having a municipality. It is primarily an administrative and commercial town surrounded by two small satellite towns—Hirakud and Burla—within a distance of 10 km.

In 1994, the existing units out of the sample fifty-eight units surveyed during 1985 were resurveyed. Only thirty-five units were found to be continuing their operations during the resurvey. The units covered by both the surveys were classified into three categories of micro business, petty production unit and small capitalist producer unit with the purpose of redefining the concept of IMS.

Results of the Comparative Analysis

Mobility and Graduation

Out of the eleven micro business units in existence in 1985, only six were in operation in 1994. Similarly, the number of petty producers declined from eighteen to ten and that of small capitalist producers from twenty-nine to nineteen during the nine-year period between 1985 and 1994. However, there had been some mobility within IMS between categories. The mortality was the highest in micro business (55 per cent) and the lowest in small capitalist production (34 per cent). Thus, the higher the status, the lower the mortality rate. On the other hand, the lower the status, the higher the mobility to upper status. Noticeably, none of the surviving thirty-five units had graduated to FMS.

The high mortality of the IMS units shows its temporary viability. Various factors such as the inability to market goods due to lower quality and higher cost, the non-availability of proper institutional help, wrong choice of project, and the unsatisfactory role of the District Industries Centre were responsible for this situation.

Characteristics of Informal Manufacturing Sector

Access to Public Utilities: During the nine-year period, IMS in general improved its access to public utilities like electricity. The proportion of units having a permanent structure decreased in the case of micro business owners and petty producers, and increased in the case of small capitalist producers. The proportion of petty producers and small capitalist producers having their own premises increased while that of micro business declined.

Expansion or Contraction: In general, there was a vertical contraction of the IMS units in Sambalpur town. In other words, the average size of employment of IMS as a whole decreased from 3.93 in 1985 to 3.31 in 1994. The same was the case with individual groups like petty producers and small capitalist producers. Despite this, the number of commercial establishments registered under the Shops and Commercial Establishments Act with the office of the Assistant Labour Commissioner, Sambalpur, increased from 184 in 1985 to 379 in 1994.

It is observed by the study (Ibid.) that the units were not employing the required number of labourers because of the increase in the bargaining power of the workers due to the increased awareness of individual rights, and the hike in the minimum wage of labour implemented by the government of Orissa. Relatively bigger units in IMS were inclined to reduce the strength of employees and to recruit multi-skill workers. As such, there was no permanence and regularity of employment in these establishments, and the owner-operators tried to evade payment of provident fund, bonus, leave compensation, etc. by resorting to various manipulative practices. These problems of workers were peculiar to small capitalist producers employing wage workers. IMS being heterogeneous, any policy measure taken to assist it, should take into account this aspect of wage payment and regularity of employment. Keeping these points in view, it becomes obvious that the classification of IMS or small-scale industries on the basis of labour composition is very relevant.

Employment Structure: First, it has been observed that the proportion of females to total participants in IMS increased mainly due to their higher participation in petty and small capitalist producer units. Second, during this period, the proportion of migrant workers drastically decreased from 57 per cent to 24 per cent. In petty production units, during the two points of time, the proportion of migrant workers remained more or less the same at around 37 per cent, whereas it decreased in small capitalist production establishments from around 67 to 23 per cent. Surprisingly, there was no micro business managed by migrant workers in 1994, though their proportion was found to be around 22 per cent during the 1985 benchmark study. The total absence of migrant workers on the micro business scene in 1994 may be due to the fact that they came to the town after having been pushed out of their villages, did business for a temporary period and returned either when the situation in their native place improved or when their business in Sambalpur was not found to be viable. They may also have gone to some other towns. With respect to assistance from the government or government agencies, they might have been discriminated against in preference to the natives by the local authorities. Since capital investment was comparatively low at around Rs. 1.33 thousand per unit in micro business, it must have been convenient for them to close down their units.

On the other hand, when the size of the unit decreases, it may be convenient on the part of small capitalist producers to retrench the migrant workers earlier than the native workers who have some political and social clout. Third, the resurvey revealed that the proportion of Hindu workers increased across the three groups. The second and third observations have led to the conclusion that there has been a localisation of informal manufacturing sector in Sambalpur town.

Performance of IMS during 1985-94

The performance of IMS consisting of fifty-eight sample units in 1985 in Sambalpur town is compared with that of the existing thirty-five IMS units in 1994. The 1985 results, wherever necessary, are suitably adjusted to the 1994 prices to make the comparison meaningful.

Small capitalist producers were found to be distinctively different from micro businesses and petty producers in many respects, such as size of employment, annual turnover, fixed capital and assets, output-fixed capital ratio and gross revenue. The performance of the former during the two points of time was better than that of micro businesses and petty producer firms in most of these areas. This may be due to the comparatively higher amount of capital investment and consequently higher level of turnover in the former. Due to higher capital investment and higher turnover, unlike micro businesses and petty producer firms, small capitalist producer firms employed wage labourers. As pointed earlier, the element of 'wage worker' was peculiar to small capitalist producer firms in our classification of IMS.

However, we cannot conclude that the higher the status, the better the performances since, on some counts, micro businesses were also observed to be performing better than petty producers. Perhaps there was some element of underemployment in production units due to the use of family labour.

IMS: Problems and Prospects

There are mainly two types of constraints faced by informal manufacturing sector: (a) internal (such as the lack of entrepreneurial or managerial skill), and (b) external (such as difficulties in accessing the market, credit, technology, raw materials and the benefits of government policy).

In the 1985 survey, IMS was found to be constrained more by external factors than by internal ones. The situation appeared to have been the same even during the follow-up survey. Shortage of raw materials and finance, for both fixed and working capital, appeared to be the major problems faced by the units. The other problems, faced especially by the small capitalist producers, were the lack of modern machinery, shortage of skilled labour, rise in the price of raw materials and inputs, and the lack of positive assistance from government agencies.

Interestingly, none of the micro business owners and petty producers were aware of the industrial policies of the government of Orissa announced during 1980, 1986 and 1992. Only three of the small capitalist producers were aware of these policies. Of these, one got a capital investment subsidy for a factory shed and the supply of scarce raw materials. Most of its products were also sold to government departments and public sector undertakings. The assets and fixed capital investment of this lone producer who got all this assistance were far higher than the other small capitalist producers. This unit also seemed to enjoy political backing and bureaucratic support. Of the two initial partners of the unit, one was the secretary of the Sambalpur town unit of a leading political party and the other a relative of a high official of the District Industries Centre, Sambalpur.

Our analysis thus points to the uniqueness of small capitalist producers among all the participating groups of IMS. This may be due to higher investment on fixed capital and assets by small capitalist producers, and employment of skilled wage workers as against the underemployment of family labour in petty production units or micro businesses.

Towards Redefining the Concept of IMS

Proposition One: Classification Based on Size

As a first step towards refining the concept of IMS, all micro businesses and petty producer firms, irrespective of their size, can be included in IMS since they do not employ any wage worker. The problem arises with respect to small capitalist producers. The question is whether they can be included in IMS. In order to answer this we have tried to bring a further differentiation among small capitalist producers (SCP) on the basis of the proportion of wage workers used. Three groups have been identified:

(a) SCP where $F>W$;
(b) SCP where $F=W$; and
(c) SCP where $F<W$.

Here,
F = family labour, whether paid or unpaid, and
W = wage workers/hired workers.

Given these, one proposition could be to eliminate small capitalist producer firms where $F>W$ and $F=W$ (first and second groups) in IMS. Out of the nineteen SCPs existing in 1994, only one had $F>W$, two had $F=W$ and the remaining sixteen had $F<W$. This means that if proposition one is accepted, sixteen out of nineteen firms otherwise having various informal characteristics will have to be excluded from IMS. Hence there is a need to bring out some other differentiation to define IMS without introducing the size criterion.

Proposition Two: Classification Based on Raw Material Used

Based on the raw materials used, IMS can be classified into the existing thirty-five IMS units of 1994 into ten categories listed earlier. Classified thus, when IMS in general of 1985 is compared with that of 1994, two out of the ten groups show some distinguishing features in terms of characteristics and performance in both the years. These two groups are tobacco-based units (*biri* only) and paint and chemical units. The former are fully linked to formal manufacturing sector through output and input. The very existence of paint and chemical unit is due to assistance from formal (government) sector in the form of support for land, subsidy for power, shed and capital investment, supply of scarce raw materials and inputs, arrangement of working capital from financial institutions and others. Thus, these two groups of units, apart from being classified on the basis of raw materials used, can also be classified under general categories, such as:

(i) Units which have some sort of written/unwritten subcontracting arrangement with a FS unit in manufacturing or other activities for the supply of 'whole' output. *Biri* units in our sample fall under this category.

(ii) Units which started initially with the assistance of the government and government agencies and whose very existence is due to government assistance for infrastructure, subsidy for investment on capital and factory sheds, supply of scarce raw materials and inputs, and the sale of output. In this sense, they are very much linked to formal and government sector. Paint and chemical units come under this category.

In the first category, there is a special type of subcontracting which may be termed 'tied subcontracting'. Tied subcontractors are fully dependent on their parent firms for their market. The *biri* roller in a *biri* unit in IMS is just like an 'out-worker' of a formal sector company working for a piece wage. Thus, the capitalist domestic industry or the activities of 'out-workers' in a FS unit may not be included in IMS. They can be treated as quasi-formal units.

In the second category, units come into existence with government assistance. Most of their products are sold to the government and/or government undertakings. Therefore, they can be considered to be quasi-formal units working in a 'protective economic environment', and hence may not be included in IMS irrespective of their size.

Thus, according to proposition two, allowance should be made for additional features of IMS such as:

(i) No written/oral contract of the unit with FS for the supply of 'whole' raw materials and inputs, and the sale of 'whole' output. That is, the unit is not a capitalist domestic industry and the entrepreneur is not an 'out-worker' of a FS unit.

(ii) Starting of manufacturing activities with own capital of the entrepreneur without any link or major assistance from the government or public institutional system. That is, the unit has started and is working in an 'unprotected economic environment'.

Initially, petty producer was defined as one who works along with paid or unpaid family labourers, the total number of workers being ten or less. Since there is no wage worker employed, there is no need to limit the size of employment of petty production establishments. Hence all micro businesses and petty producer firms can be included in IMS irrespective of their size.

Similarly, small capitalist producer (SCP) firms were defined in the beginning as those which employ hired workers with or without paid/unpaid family labour, the total workers being ten or less. Since in more then 80 per cent of small capitalist producer units, the number of family labour (F) was less than wage workers (W), the proposition to eliminate the small capitalist producer units where F<W from IMS does not seem relevant. The limit of ten is also irrelevant since some of the small capitalist producers are working informally, even though the number of workers is more than ten, say for instance in a brick kiln. After considering all these we propose that along with SCPs employing up to ten wage/hired workers with or without family labour, those SCP units with total workers exceeding ten (including wage/hired workers and family labour) can be included in IMS, provided they satisfy the following conditions.

(i) *Unprotected wage work*: There is no written or oral contract between the employer and workers which is effective in a court of law or judiciary. Wage work is unprotected irrespective of whether workers are regular, irregular, short term, or long term. There is no effective trade union of workers or any effective government assistance to employees. Since the very existence and continuity of the units in IMS are uncertain, the tenure of workers is entirely unprotected. Wage is determined on the basis of bargaining power, and there is no effective role of institutions in the fixation of wage. The unit is under no wage legislation or any 'effective' institutional legislation.

(ii) *Unprotected economic environment*: Units are set up with the entrepreneurs' own funds and effort and without any major assistance from the public institutional system or government agencies. That is, units have been working in an 'unprotected' economic environment.

The forementioned arguments suggest that irrespective of the size of the enterprise, the participants or labour composition in different categories of IMS is as follows:

(a) micro businesses (*E*);
(b) petty producer units (*E+F*); and
(c) small capitalist producer firms (*E+F+W* or *E+W*), who have 'unprotected' wage work, and an 'unprotected' economic environment,

where
E = the entrepreneur
F = Fp+ Fu= family labour
Fp= paid family labour
Fu= unpaid family labour
W = wage/hired workers

Policy Implications

Different types of assistance are provided to small-scale units by the public institutional system with the implicit objective of employment generation. It is observed in this study that in spite of various industrial policies of the government of Orissa, very small manufacturing units in the IMS, such as micro businesses and petty producer units, were unable to get any assistance during the nine-year period 1985-94. At the same time, the average size of employment in these units has increased. On the other hand, small capitalist producers (comparatively larger units) were able to get various benefits, though their average size of employment was declining. Assistance to small-scale units in the name of employment generation, without making any distinction between different types of firms, appears to be irrelevant.

Small-scale units use various tactics, like inflating capital costs or changing the name of the unit at certain intervals to be 'new', to get more concessions and subsidies. Some of the entrepreneurs are able to procure these concessions through political backing and bureaucratic support. Most of them vanish after getting subsidies. Sometimes they convert factory sheds, if any, to a residential complex.

Instead of creating a new 'parasite middle class' through political and bureaucratic support, it is better to assist the 'existing and operating' informal manufacturing units to meet their requirements relating to credit, market and so on. They need adequate and timely credit, but not necessarily at a concessional rate of interest. The IMS units may even be ready to pay a higher than normal market rate of interest since they usually procure their finances at hefty rate of interest in informal credit market.

There is a need for 'institutional radicalisation' with a thrust on credit, access to land and other assets, and improved technology, among other things (Mobogunje: 1991). There is also a need for re-examination of the role of IMS in the economy. Any industrial policy for the small-scale and informal manufacturing sectors must seek to assist the 'productive' entrepreneur while discouraging the 'rent-seeking' ones in the parasite middle class.

It is argued that there is a need to reorient economic policies to promote more and better employment in order to reduce poverty and increase the representative voice of workers—especially informal workers—in the processes and institutions that determine economic policies (Chen, Vanek and Heintz: 2006).

Pre- and Post-Reform Growth

A study similar to our, but on the basis of macro-data in India observed that the growth of IMS in India over the period 1984 to 2000 is neither smooth nor uniform (D. Mukherjee: 2004). It argued that IMS cannot be labelled either a distress driven sink or a dynamic alternative economic avenue in blanket terms as existence of both the segments is detected. According to the study (Ibid.), conceptually, IMS in India is generally taken to be composed of three segments—OAMEs, NDMEs and DMEs. Own account manufacturing enterprises (OAMEs) operate with no hired workers employed on a fairly regular basis. Non-discretory manufacturing establishments (NDMEs) employ less than six workers including household workers. Directory manufacturing establishments (DMEs) employ six or more workers with at least one hired worker but not registered under the Factory Act.

In the pre-reform period in India, number of enterprises in both OAMEs and NDMEs decreased while that of DMEs increased. Employment decreased in the OAMEs and rural NDMEs but increased in the DMEs and urban NDMEs. The decreasing trend of number of enterprises and employment continued during 1989-94 transition period. There was a reversal of trend during the post-reform period. While employment increased in all the segments of IMS, number of enterprises increased for all the segments except rural NDMEs and rural DMEs. Considering the heterogeneity in IMS, the study (Ibid.) recommended that distinctly different set of policies be provided for different segments of IMS.

6

Summary and Conclusion

There was an upsurge of debate, discussion and controversy around the concept of informal sector during 1970s and 1980s particularly in the context of developing countries which have experienced a pattern of urbanisation with a large part of labour force in low-income activities. The use of informal-formal sector dichotomy was used as a conventional method of analysing the structure of urban economies. In spite of various criticisms, difficulties, confusion and deficiencies in the concept of IS at the initial stage, there was a rapid diffusion of the concept since 1971 by the leading international institutions such as ILO, UNO, IBRD, etc. and several scholars.

Concept and Dynamics

The concept of IS which had elements of vagueness and overgeneralisation has further been refined by several scholars by adding some more characteristics for distinguishing it from FS on the basis of their relation to state, participants, structure, mode of production, linkages, growth, size, productivity, income, etc. Gradually, there is emergence of clarity on definition of IS. The 15th International Conference of Labour Statisticians (ICLS), after reviewing the ILO background report, clarified the concept of IS by specifying what is considered to be principal and distinctive features of IS enterprises. An enterprise is classified as informal if it meets any (or all) of the three criteria: (i) employs less than 10 persons, (ii) not registered, and (iii) not maintain complete accounts. In one of our studies at two points of time, we have also refined the concept and included (i) micro-business, (ii) petty producers, and (iii) small capitalist producers having (a) 'unprotected' wage work and (b) 'unprotected' economic environment, in IS. Our categorisation helped us in our analysis and in formulating policy suggestions. In the era of liberalisation and privatisation, the significance of the concept of IS, instead of marginalised, is gaining prominence due to shift of labourers from formal to informal sector which is acting as safety net.

Rural Non-Farm Sector

Since the agricultural sector in the rural areas is oversaturated, the growth of RNFS is seen as a promising sector to play a positive role in promoting growth and welfare. Therefore, there is a case for focusing on RNFS from the point of view of income distribution. RNFS is defined as the sector which includes all economic activities in rural areas except agriculture, livestock, fishing and hunting. It is not a homogeneous sector. RNFS includes a wide range of economic activities whose composition may vary from country to country and, within a country from region to region, from state to state.

The three past well-known hypotheses argue that the level of rural non-farm sector is either determined by agricultural development, or urbanisation or forced expulsion of labourers out of agriculture. It is imperative to think of dualism in RNFS. RNFS can be classified as formal and informal, and a further distinction can be made between traditional and modern within informal RNFS. In our analysis of RNFS, we come to conclusion that more even distribution of land and rural assets including water from irrigation and equal access to common property resources as distinct from subsidisation of mechanisation of agriculture and, expansion of primary and secondary education rather than subsidisation of higher education, may help the growth and expansion of informal RNFS activities by increasing productivity and skill of workers, rural income and improving distribution of such income. Thus, more even distribution of land and rural assets, as also literacy and education are the two most important factors determining the level and growth of RNFS activities, particularly the informal component of it.

Linkages

The linkages between IS and FS are heterogeneous. Under integrated condition, in most of the cases, IS is exploited by FS through latter's subordination under a mechanism whereby FS appropriates the surplus generated in IS. It is observed that there were more linkages between the two sectors in India through subcontracting in the engineering as well as automobile industries. The formal sector establishments did not provide any financial assistance to their sub-contractors and SSI/IS units. Though, the small and IS units received technical guidance from the parent firms, technology transfer was very limited.

In most of the cases, the forward linkages of IS with FS was negligible but backward linkage—both direct and indirect—was prevalent. However, whatever may be the linkages, always IS was at disadvantage due to its weak bargaining power. In general, IS was exploited by FS either through input or output linkages. There were complex set of factors which influence sub-contracting out of which, saving labour cost and capital cost was prominent. As the industrial development proceeds, there is transformation of linkages and sub-contracting arrangements including re-sub-contracting chain. Iron and Steel industry is believed to have the highest total linkages which may not be found in case of aluminium industry. From our studies in Orissa, the same trend is observed that the linkage—both forward and backward—of NALCO are very negligible in Orissa in comparison to RSP in Rourkela.

Rural-Urban Migration

The most popular model is that of Todaro which regards the migration as a two-stage phenomenon by recognising a dichotomy in an urban economy. Various empirical studies show that migration does occur from low to high income regions but some debate continues regarding the relative importance of the characteristics of origin and destination in explaining migration. Various independent variables are used to explain as determinants of rural-to-urban migration such as level of education, personal contacts, geographical distance, bright lights, government policies, risk avoidance, urban-rural income difference, agricultural unemployment under-employment and others.

The relationship between migration and labour force participation in the towns and cities is indeterminate, as found in both theoretical and empirical literature. The scattered empirical studies do not support the contention that rural-to-urban migration may be limited to marginal employment in IS.

The studies on rural-to-urban migration in Orissa confirmed the close link between UIS and rural-to-urban migration; and found push factors, in general, as more important determinant than pull factors in the process of rural-to-urban migration. However, pull factors are predominant in emerging industrial centres for IS entrepreneurs and in coal mine region for most of the IS workers. Besides graduation from IS to FS, there was also backward movement of IS workers to their native places. Distance plays a role in rural-to-urban migration. The contention that only the poorest of the poor in rural areas migrate to towns and cities is not necessarily true. It is not only the worse-off but also the better-off households in the villages, who migrate to urban centres in search of jobs.

Policy Implication

Assistance to IS units in the name of employment generation, without making any distinction between different types of firms, appears to be irrevalent. There is a need for bringing distinction of different categories of IS for having any policy to assist IS. Morever, any policy measure taken to assist it should take into account various aspects of wage payments, regularity of employment, payment of providend fund, bonus, leave concession, etc. of the IS units. Instead of creating a new 'parasite middle class' through political and bureaucratic support, it is better to assist the 'existing and operating' informal units to meet their requirements relating to credit, marketing and so on. They need adequate and timely credit, but not necessarily at a concessional rate of interest. Any policy for informal sector must seek to assist the 'productive' entrepreneur while discouraging 'rent-seeking' ones.

More even distribution of land and rural assets, as also literacy and education are the two most important factors determining the level and growth of rural non-farm sector activities, particularly the informal component of it. The equity hypothesis is more plausible than the inter-linkage hypothesis and urbanisation hypothesis. Ancillarisation through government programme of Ancillary Development Scheme may help in establishing linkages between IS and FS. But some sector like steel and iron has more linkage effects than aluminium industry. Moreover, some policies like Freight Equalisation Policy since independence upto 1994 were against the development of linked units around the big FS units like steel plant.

The social inequality of rural areas continues to surface in urban informal sector in the form of division of labour among the rural-urban migrant workers. However, there seems to be a reduction of income inequality among them due to migration. Thus, rural-to-urban migration as one of the tools to alleviate rural poverty may not easily be challenged; more so, in the era of liberalisation and globalisation, when large number of workers are joining IS being driven out from FS including both public and private sectors. To alleviate rural poverty and unemployment, the growth of RNFS is also desirable.

References

Akinboade, Oludele Akinloye (2005); "A Review of Women Poverty and Informal Trade Issues in Eastern and Southern Africa", *International Social Science Journal*, 184, Vol. 57, No. 2, pp. 255-275.

Alexander, P. et al. (eds.) (1991); *In the Shadow of Agriculture: Non-Farm Activities in the Javanese Economy: Past and Present*, Royal Tropical Institute, Amsterdam.

Amin, A. T. M. Nurul (1996); "Editorial Introduction: The Informal Sector Paradigm: Analytical Contributions and Development Role", *Regional Development Dialogue*, Vol. 17, No. 1, Spring, pp. vi-xxvii.

Amin, Samir (ed.) (1974); *Modern Migration in Western Africa*, OUP, London.

Amin, Samir (ed.) (1983); *Human Resources, Employment and Development, Vol. 5: Developing Countries*, Macmillan, London.

Annavajhula, J. C. B. (1989); "Japanese Sub-contracting System", *Economic and Political Weekly*, Vol. 24, No. 8, February 25, pp. M 15-23.

Aziz, Abdul (1984); *Urban Poor and Urban Informal Sector*, Ashish Publishing House, New Delhi.

Bagachwa, M. D. and F. Stewart (1992); "Rural Industries and Rural Linkages", in F. Stewart, S. Lall and S. Wangwe (eds.), *Alternative Development Strategies in Sub-Saharan Africa*, Macmillan Press Ltd., London.

Balkrishna, Suresh and K. Gopalkumar (1995); "Rural Enterprises in a Market Economy: The Emerging Agenda", *Man and Development*, Vol. 17, No. 3, September, pp. 61-68.

Banerjee, Biswajit (1983); "The Role of Informal Sector in the Migration Process: A Test of Probabilistic Migration Models and Labour Market Segmentation for India", *Oxford Economic Papers* (New Series), Vol. 35, No. 3, pp. 399-422.

Basant, Rakesh (1993); "Diversification of Economic Activities in Rural Gujarat: Key Results of a Primary Survey", *The Indian Journal of Labour Economics*, Vol. 36, No. 3, July-September, pp. 361-86.

Basant, Rakesh (1994); "Economic Diversification in Rural Areas: Review of Processes with Special Reference to Gujarat", *Economic and Political Weekly*, Vol. 29, No. 39, September 24, pp. A107-16.

Basant, Rakesh and B.L. Kumar (1994); "Rural Non-Agricultural Employment in India: A Review of Available Evidence", in Pravin Visaria and Rakesh Basant (eds.), *Non-Agricultural Employment in India: Trends and Prospects*, Sage Publication India Pvt. Ltd., also in *Social Scientist*, Vol. 17, Nos. 1-2, January-February, 1989.

Bhalla, Sheila (1990); "Employment in Indian Agriculture: Retrospect and Prospect", *Social Scientist*, Nos. 192 and 193, pp. 3-21.

Bhalla, Sheila (1993); "Trends of Some Proposition About the Dynamics of Changes in the Rural Workforce Structure", *Indian Journal of Labour Economics*, Vol. 36, No. 3, July-September, pp. 428-39.

Bharadwaj, Krishna (1979); "Towards a Macro-Economic Framework for a Developing Economy: The Indian Case", *The Manchester School of Economics and Social Studies*, September, No. 3, pp. 270-302.

Bhattacharya, Prabir (1998); "The Informal Sector and Rural-to-Urban Migration: Some Indian Evidence", *Economic and Political Weekly*, Vol. 33, No. 21, May 23, pp. 1255-62.

Binswanger, Hans P. and Jauchim Von-Braun (1991); "Technological Change and Commercialisation in Agriculture: The Effect on the Poor", *The World Bank Research Observer*, Vol. 6, No. 1, January, pp. 57-80.

Birbeck, Chris (1978); "Self-employed Proletarians in an Informal Factory: The Case of Cali's Garbage Dump", *World Development*, Vol. 6, No. 9/10, September/October, pp. 1173-85.

Boeke, J. H. (1953); *Economics and Economic Policy of Dual Societies: As Exemplified by Indonesia*, International Secretariat, Institute of Pacific Relations, New York.

Bombay Chamber of Commerce and Industry (1979); *Role of Organised Industry in the Effective Development of Small-Scale Industry*, Bombay.

Bose, A. N. (1978); *Calcutta and Rural Bengal: Small Sector Symbiosis*, Minerva, Calcutta.

Bose, A. N. (1980); "Informal Sector: Problems and Prospects", *Nagarlok*, Vol. 12, No. 2, April-June, pp. 24-34.

Breman, Jan (1976); "A Dualistic Labour System? A Critique of the 'Informal Sector' Concept—I, II, III, "*Economic and Political Weekly*, Vol. 11, Nos. 48, 49 and 50, November 27, December 4, 11, 1976, pp. 1870-76, 1905-8, and 1993-2004.

Breman, Jan (1977); "Labour Relations in 'Formal' and 'Informal' Sector: Report of a Case Study in South Gujarat", *The Journal of Peasant Studies*, Vol. 4, Nos. 3 and 4, April and July, pp. 171-205 and 337-359.

Bromley, Ray (1978a); "Introduction: The Urban Informal Sector: Why is it Worth Discussing" '*World Development*, Vol. 6, No. 9/10, September/October, pp. 1033-39.

Bromley, Ray (1978b); "Organisation, Regulation and Exploitation in the So Called 'Urban Informal Sector': The Street Traders of Cali, Columbia", *World Development*, Vol. 6, No. 9/10, September/October, pp. 1161-71

Chadha, G.K. (1992); *Non-Farm Sector in Rural China: Trends and Variation in Labour Productivity*, ILO/ARTEP Working Paper, New Delhi, May.

Chadha, G.K. (1993); "Non-Farm Employment for Rural Households: Evidence and Prognosis", *Indian Journal of Labour Economics*, Vol. 38, No. 3, July-September, pp. 296-327.

Chandrasekhar, C. P. (1993); "Agrarian Change and Occupational Diversification: Non-Agricultural Employment and Rural Development in West Bengal", *Journal of Peasant Studies*, Vol. 20, No. 2, pp. 205-70.

Charms, Jacques (1996); "Progress in Measurement of Informal Sector Employment", *Regional Development Dialogue*, Vol. 17, No. 1, Spring, pp. 18-30.

Chen, Martha, Joann Vanek and Jomes Heinz (2006); "Informality, Gender and Poverty: A Global Picture", *Economic and Political Weekly*, Vol. 41, No. 21, May 27, pp. 2131-2139.

Choudhury, T. D. (1989); "A Theoretical Analysis of the Informal Sector", *World Development*, Vol. 17, No. 3, January-June, pp. 351-355.

Choudhury, Utpal (2000); "An Alternative Approach to Informal Sector", *Asian Economic Review*, Vol. 42, No. 3, pp. 470-479.

Collier, P. and D. Lall (1986); *Labour and Poverty in Kenya, 1900-1980*, Clarendon Press, Oxford.

Das, Keshab (2000); "Workers and Earnings in Informal Manufacturing: Evidence and Issues in Estimation", *The Indian Journal of Labour Economics*, Vol. 43, No. 2, April-June, pp. 261-276.

Dasgupta, Biplab (1973); "Calcutta's Informal Sector", *Bulletin: Institute of Development Studies*, Vol. 5, October, pp. 53-75.

Dasgupta, Sukti (2003); "Structural and Behavioural Characteristics of Informal Service Employment: Evidence from a Survey in New Delhi", *The Journal of Development Studies*, Vol. 39, No. 3, February, pp. 51-80.

Davies, R. (1979); "Informal Sector or Subordinate Mode of Production? A Model" in R. Bromley and C. Gerry (eds.), *Casual Work and Poverty in the Third World Cities*, John Wiley and Sons.

Dev, S. Mahendra (1990); "Non-Agricultural Employment in Rural India: Evidence at a Disaggregate Level", *Economic and Political Weekly*, Vol. 25, No. 28, July14, pp. 1526-36.

Dhesi, Atur S. and Urvashi Wadhwa (1983); "The Informal Sector in Nangel (Punjab): An Overview", *Indian Journal of Regional Science,* Vol. 110, No. 3, September, pp. 199-218.

Duraisamy, P. and S. Narasimhan (2000); "Migration, Remittances and Family Ties in Urban Informal Sector", *The Indian Journal of Labour Economics,* Vol. 43, No. 1, January-March, pp. 111-119.

Durgaprasad, P. (1995); "Paradigms for Action and Development Alternatives: The Case of Rural Industrialisation in India", *Man and Development,* Vol. 17, No. 3, September.

Eapen, Mridul (1994); "Rural Non-Agricultural Employment in Kerala: Some Emerging Tendencies", *Economic and Political Weekly,* Vol. 29, No. 21, May 21, pp. 634-53.

Eapen, Mridul (1995); "Rural Non-Agricultural Employment in Kerala: Inter-District Variation", *Economic and Political Weekly,* Vol. 30, No. 12, March 25, pp. 634-53.

Economic Times, The, Kolkatta, October 30, 1990.

Engles, F. (1973); "The Housing Question", *Marx and Engels: Selected Works,* Vol. 2, Progress Publishers, Moscow, pp. 295-375.

Evans, H. and P. Ngau (1991); "Rural-Urban Relationship, Household Income Diversification and Agricultural Productivity", *Development and Change,* Vol. 22, pp. 519-45.

Fabella, R. (1986); "Rural Non-Farm Employment in the Philippines: Composition, Growth and Seasonality", in R. T. Shand (ed.), *Off-Farm Employment in the Development of Rural Asia,* Australian National University, Canberra.

Fei, John C. H. and Gustav Ranis (1967); *Development of the Labour Surplus Economy: Theory and Policy,* Richard D. Irwin, Inc, Illinois.

Geertz, Clifford (1963); *Peddlers and Princes: Social Change and Economic Modernisation in Two Indonesian Towns,* The University of Chicago Press.

Gerry, Chris (1978); "Petty Production and Capitalist Production in Dakar: The Crisis of Self-employment",

World Development, Vol. 6, No. 9/10, September/October, pp. 1147-60.

Gerry, Chris (1979); "Small-scale Manufacturing and Repairs in Dakar: A Survey of Market Relations within the Urban Economy" in R. Bromley and Chris Gerry (eds.), *Casual Workers and Poverty in Third World Cities*, John Wiley and Sons.

Godfrey, Martin (1979); "Rural-Urban Migration in a 'Lewis-Model' Context", *The Manchester School of Economics and Social Studies*, September, No. 3, pp. 230-47.

Government of India (1993); *Selected Socio-Economic Statistics for India*, Central Statistical Organisation, New Delhi.

Government of India, *Census of India* 1961, 1971, 1981, 1991, 2001.

Government of India; National Commission for Enterprises in the Unorganised Sector (2006); *Social Security for Unorganised Workers: Report*, New Delhi.

Government of Orissa (1981); *Report of a Socio-Economic Study in Hirakud Ayacut Area*, Bureau of Statistics and Economics, Bhubaneswar.

Government of Orissa (1985); *Statistical Abstract of Orissa*, Bureau of Statistics and Economics, Bhubaneswar and various other publications.

Government of Orissa (2001); *Economic Survey*, and various other issues.

Government of Pakistan (1994); Ministry of Manpower and Overseas Pakistanis and Growth Dynamics University Institute, Eramus University, Roterdam; *Informal Sector Employment in Pakistan: New Approaches and New Insights*, Chapter-2, "Measurement of Informal Sector Employment: Conceptual and Definitional Issues", pp. 9-20, Islamabad.

Greenwood, Michael J. (1975); "Research on Internal Migration in the United States: A Survey", *Journal of Economic Literature*, Vol. 13, No. 2, pp. 397-433.

Guisinger, Stephen and Mohammed Irfan (1980); "Pakistan's Informal Sector", *The Journal of Development Studies*, Vol. 16, No. 4, July, pp. 412-26.

Gupta, L. C. (1990); *Dynamic of Regional Growth Process: Basic Proposition and Case History*, IDBI, Mumbai.

Haan, Arjan De (1999); "Livelihoods and Poverty: The Role of Migration—A Critical Review of the Migration Literature", *The Journal of Development Studies*, Vol. 36, No. 2, December, pp. 1-47.

Haan, Arjan De (2002); "Migration and Livelihood in Historical Perspective: A Case Study of Bihar, India", *The Journal of Development Studies*, Vol. 35, No. 5, June, pp. 115-42.

Haan, Arjan De and Ben Rogaly (2002); "Introduction: Migrant Workers and their Role in Rural Change", *The Journal of Development Studies*, Vol. 35, No. 5, June, pp. 1-14.

Haggblade, S., P. Hazell and J. Brown (1989); "Farm-Non-Farm Linkages in Rural Sub-Saharan Africa", *World Development*, Vol. 17, Nos. 8 and 13, pp. 1173-1201.

Hakam, N. Ali (1983); "The Apprentice System in Ghana: An Instrument for Technology Transfer and Diffusion" in S. Watanabe (ed.), *Technology, Marketing and Industrialisation: Linkages Between Large and Small Enterprises*, Macmillan, New Delhi, 1983.

Harberger, A. C. (1971); "On Measuring the Social Opportunity Cost of Labour", *International Labour Review*, Vol. 103, No. 6, June, pp. 559-79.

Harriss, Barbara (1978); "Quasi-Formal Employment Structure and Behaviour in the Unorganised Urban Economy and the Reverse: Some Evidence From South India", *World Development*, Vol. 6, No. 9/10, September/October, pp. 1077-86.

Harriss, Barbara (1987); "Regional Growth Linkages From Agriculture and Resource Flows in Non-Farm Economy", *Economic and Political Weekly*, Vol. 22, Nos.1 and 2, January 19, pp. 31-46.

Harriss, John (1982); "Character of an Urban Economy: Small-scale Production and Labour Market in Coimbatore", *Economic and Political Weekly*, Vol. 17, Nos. 23 and 24, June 5, 12, pp. 945-54 and 993-1002.

Harriss, John, K. P. Kannan and Gerry Rodgers (1990); *Urban Labour Market Structure and Job Access in India: A Study of Coimbatore*, ILO, Geneva.

Hart, Keith (1973); "Informal Income Opportunities and Urban Employment in Ghana", *The Journal of Modern African Studies*, Vol. 11, No. 1, pp. 61-89.

Hayami, Yujiro, A. K. Dikshit and S. N. Mishra (2006); "Waste Pickers and Collectors in Delhi: Poverty and Environment in an Urban Informal Sector", *Journal of Development Studies*, Vol. 42, No. 1, January, pp. 41-69.

Hazell, P. et al. (1991); "Economic Changes Among Village Households", in P. Hazell and C. Ramaswamy (eds.), *The Green Revolution Reconsidered: The Impact of High Yielding Rice Varieties in South India*, John Hopkins University Press, Batimore, Md.

Hazell, P. and S. Haggblade (1990); *Rural-Urban Growth Linkages in India*, PRE Working Paper Series No. 430, ARDD, World Bank, Washington.

Higgins, Benjamin (1966); *Economic Development: Principles, Problems and Policies*, Central Book Depot, Allahabad.

Hilhorst, J. G. M. (1984); "City Size and the Informal Sector", *Regional Development Dialogue*, Vol. 5, No. 2, pp. 86-95.

Hirschman, Albert O. (1958); *The Strategy of Economic Development*, Yale University Press, New Haven.

Hossain, M. (1987); "Employment Generation Through Cottage Industries—Potential and Constraints: The Case of Bangladesh", in R. Islam (eds.), *Rural Industrialisation and Employment in Asia*, ILO (ARTEP), New Delhi.

Hussmanns, Raff (1995); "ILO's Recommendation on Methodologies Concerning Informal Sector Data Collection", in B. Herman (ed.), *Papers, Proceedings and Recommendations of the International Seminar on Informal Sector Employment Statistics*, Ministry of Manpower and Overseas Pakistanis, Government of Pakistan and Growth Dynamics University Institute, Eramus University of Roterdam, pp. 47-62.

Hymer, S. and S. Resnick (1969); "A Model of an Agrarian Economy with Non-Agricultural Activities", *American Economic Review*, Vol. 50, pp. 493-500.

IMF Survey (1995a); Vol. 24, No. 10, May 22.

IMF Survey (1995b); Vol. 24, No. 13, July 3 and various other issues.

International Labour Organisation (1972); *Employment, Income and Equality: A Strategy for Increasing Productive Employment in Kenya*, ILO, Geneva.

International Labour Organisation (1983); *Promotion of Employment and Income for Rural Poor Including Rural Women Through Non-Farm Activities*, International Labour Organisation, Geneva.

International Labour Organisation (1992); *Statistics of Employment in the Informal Sector*, Paper-III, Fifteenth International Conference of Labour Statisticians, Geneva, 19-28 January, 1993.

International Labour Organisation (1995); *World Employment 1995: An ILO Report*, Geneva.

Isgut, Alberto E. (2004); "Non-farm Income and Employment in Rural Honduras: Assessing the Role of Locational Factors", *The Journal of Development Studies*, Vol. 40, No. 3, February, pp. 59-86.

Jayaraj, D. (1994); "Determinants of Rural Non-Agricultural Employment", in P. Visaria and R. Basant (eds.), op. cit., Sage Publication, New Delhi.

Joshi, Heather (1980); "The Informal Urban Economy and its Boundaries", *Economic and Political Weekly*, Vol. 15, No. 13, March 29, pp. 638-44.

Joshi, Heather and Vijaya Joshi (1976); *Surplus Labour and the City: A Study of Bombay*, OUP, Delhi.

Kantor, Paula, Uma Rani and Jeemol Unni (2006); "Decent Work Deficits in Informal Economy: Case of Surat", *Economic and Political Weekly*, Vol. 41, No. 21, May 27, pp. 2089-2097.

Karanshawy, Hatem El. and Mohammad Sakar (1983); "*Linkages in the Egyptian Cotton Weaving Industry*" in S.Watanabe (ed.), op.cit.

Khan, Abdullah Abusayed (2005); "The Status of Working Women in Informal Sector: The Case of Bangaladesh", *Indian Journal of Regional Science*, Vol. 37, No. 1, pp. 111-119.

King, Kenneth J. (1974); "Kenya's Informal Machine-Makers: A Study of Small-Scale Industry in Kenya's Emergent Artisan Society", *World Development*, Vol. 2, Nos. 4 and 5, April-May, pp. 9-28.

King, Kenneth J. (1975); "Skill Acquisition in the Informal Sector in a African Economy: The Kenya Case", *The Journal of Development Studies*, Vol. 11, No. 2, January, pp. 108-22.

Kulshreshtha, A. C. (1998); "Informal Sector in India: Conceptual and Estimational Issues in the Context of the U. N. System of National Accounts", *The Indian Journal of Labour Economics*, Vol. 41, No. 3, July-September, pp. 449-58.

Kulshreshtha, A. C. and Gulab Singh (1999); "Gross Domestic Product and Employment in the Informal Sector of the Indian Economy", *The Indian Journal of Labour Economics*, Vol. 42, No. 2, April-June, pp. 217-230.

Kumar, Alok (1993); "Rural Non-Farm Employment: A Static and Dynamic Study of Inter-State Variation", *Indian Journal of Labour Economics*, Vol. 36, No. 3, July-September, pp. 450-54.

Kumar, Sanjay and N. K. Sharma (2003); "Informal Sector: An Indian Perspective", *The Indian Journal of Labour Economics*, Vol. 46, No. 1, January-March, pp. 95-108.

Kundu, Amitabh (1991); "Growth of Non-Agricultural Employment—A Hypothesis on Rural-Urban Linkages", *IASSI Quarterly*, Vol. 10, No. 2, October-December.

Kundu, Amitabh (1998); "Introduction", *The Indian Journal of Labour Economics*, (A Special Issue on Informal Sector), Vol. 41, No. 3, July-September, pp. 439-448.

Kundu, Amitabh and Alakh N. Sharma (eds.) (2001); *Informal Sector in India: Perspectives and Policies*, Institute for Human Development and Institute of Applied Manpower Research, New Delhi.

Kundu, Amitabh, N. Lalitha and Sant Lal Arora (2001); "Growth Dynamics of Informal Manufacturing Sector in Urban India: An Analysis of Interdependencies", in Amitabh Kundu and Alakh N. Sharma (eds), op. cit.

Lal, Sanjoy (1980); "Vertical Inter-Firm Linkages in LDC: An Empirical Study", *Oxford Bulletin of Economics and Statistics*, Vol. 42, No. 3.

Lanjouw, Jean O. and Peter Lanjouw (1995); *Rural Non-Farm Employment: A Study*, Policy Research Working Paper 1463, World Bank, Washington.

Larsen and Toubro Limited (1990); *L &T 90, A New Decade Dawns.*

Lenin, V. I. (1977); "The Development of Capitalism in Russia", *Collected Works*, Vol. 3, Progress Publishers, Moscow, pp. 21-607.

Lewis, B. and E. Thorbecke (1992); "District Level Economic Linkages in Kenya: Evidence Based on a Small Regional Social Accounting Matrix", *World Development*, Vol. 20, No. 6.

Lewis, W. Arthur (1954); "Economic Development with Unlimited Supplies of Labour", *The Manchester School of Economics and Social Studies*, May, pp. 139-91.

Lewis, W. Arthur (1979); "The Dual Economy Revisited", *The Manchester School of Economics and Social Studies*, September, No. 3, pp. 211-29.

Leys, Colin (1973); "Interpreting African Underdevelopment: Reflections on the ILO Report on Employment, Incomes and Equality in Kenya", *African Affairs*, October 1973, reprinted in G. M. Meier (ed.), *Leading Issues in Economic Development*, OUP, New York.

Liedholm, Carl and Enyinna Chuta (1979); *Rural Non-Farm Employment: Review of the State of the Art*, Michigan State University, Michigan.

Lipton, Michael (1984); "Family Fungibility and Formality: Rural Advantage of Informal Non-Farm Enterprise *versus* the Urban-Formal State", in S. Amin (ed.), *Human Resources, Employment and Development*, Vol. 5: *Developing Countries*, Macmillan, London, 1984.

Little, I. M. D. (1987); "Small Manufacturing Enterprises in Developing Countries", *World Bank Economic Review*, Vol. 1, No. 2, January, pp. 203-35.

Lubell, Harold (1974); *Urban Development and Employment: The Prospects for Calcutta*, ILO, Geneva.

Marx, Karl (1971); *Capital*: Vol. 11, Progress Publishers, Moscow.

Marx, Karl (1974); *Capital*, Vol. 1, Progress Publishers, Moscow.

Marx, Karl and Frederick Engels (1976); *The German Ideology*, Progress Publishers, Moscow.

Mathur, Ashok and P. K. Pani (1993); "The Character of Rural Industrialisation and its Employment Impact in India: An Interregional Perspective", *Indian Journal of Labour Economics*, Vol. 36, No. 3, July-September.

Mazumdar, Atreyi (1980), *In-Migration and Informal Sector: A Case Study of Urban Delhi*, Vision Books, New Delhi.

Mazumdar, D. (1973); "Labour Supply in Early Industrialisation: The Case of the Bombay Textile Industry", *The Economic History Review*, (2nd series), Vol. 26, No. 3, August, pp. 477-96.

McGee, T. G. (1973); "Peasants in the Cities: A Paradox, A Most Ingenuous Paradox", *Human Organisation*, Vol. 32, No. 2, Summer, pp. 135-42.

McGee, T. G. (1975); *The Urbanisation Process in the Third World Explorations in Search of a Theory*, G. Bells and Sons Ltd., London.

McGee, T. G. (1982); "Labour Mobility in Fragmented Labour Markets: The Role of Circulatory Migration in Rural-Urban Relations in Asia" in Helen I Safa (ed.), *Towards a Political Economy of Urbanisation in Third World Countries*, OUP, New Delhi.

Mehta, G. S. (2003); *Non-Farm Economy and Rural Development*, Anmol Publications, New Delhi.

Mellor, John W. (1976); *The New Economics of Growth—A Strategy for India and the Developing World*, Cornell University Press, Ithaka.

Mishra, S. N. (1998); *Dynamics of Rural-Urban Migration in India*, Anmol Publication, New Delhi.

Mitra, Arup (1993); "Rural Non-Farm Employment, Poverty and Women", *Indian Journal of Labour Economics*, Vol. 36, No. 4, July-September, pp. 455-69.

Mitra, Arup (1997); "Labour Flexibility in the Organised Sector: With Special Reference to Manufacturing", in IAMR (ed.), *Concept and Measurement of Labour Flexibility with Special Reference to Organised Sector in India*, Proceedings of the IAMR, New Delhi, 1997.

Mitra, Arup (2005); "Women in Urban Informal Sector: Perpetuation of Meagre Earnings", *Development and Change*, Vol. 36, No. 2, March, pp. 291-316.

Mitra, Arup (2006); "Labour Market Mobility of Low Income Households", *Economic and Political Weekly*, Vol. 41, No. 21, May 27, pp. 2123-30.

Mobogunje, A. L. (1991); "A New Paradigm for Urban Development", *Proceedings of the World Bank Annual Conference on Development Economics*, World Bank, Washington DC.

Mohapatra, Ajay (1990); *Impact of Industrialisation on the Tribal Population in the Rourkela Region of Sundargarh District, Orissa*, Doctoral Dissertation for Utkal University, Bhubaneswar.

Moser, C. O. N. (1978); "Informal Sector or Petty Commodity Production: Dualism or Dependence in Urban Development", *World Development*, Vol. 6, No. 9/10, September/October, pp. 1041-64.

Mosse, David, Sanjeev Gupta, Mona Mehata, Vidya Shah, Julia Rees and the Krib Project Team (2002); "Brokered Livelihood: Debt, Labour Migration and Dependent in

Tribal Western India", *The Journal of Development Studies*, Vol. 38, No. 5, June, pp. 59-88.

Mukherjee, Deepa (2004a); "Informal Manufacturing Sector in India: Pre- and Post-Reform Growth Dynamics", *The Indian Journal of Labour Economics*, Vol. 47, No. 2, April-June, pp. 293-310.

Mukherjee, Deepa (2004b); "Problems and Prospects of Informal Manufacturing Sector in an Industrial Economy—Lessons for Regional Development", *Indian Journal of Regional Science*, Vol. 36, No. 2, pp. 1-15.

Mukhopadhyay, Ishita (1998); "Calcutta's Informal Sector—Changing Pattern of Labour Use", *Economic and Political Weekly*, Vol. 33, No. 47-48, November 21, pp. 3075-80.

Mukhopadhyay, Swapna (1985); "Rural Non-Farm Sector in Asia: A Characterisation", *Economic and Political Weekly*, Vol. 20, No. 22, June, pp. 966-73.

Myrdal, Gunnar (1968); *Asian Drama: An Enquiry into the Poverty of Nations*, Penguin, London, Vol. 2, Chap. 23, Sec. 4, pp. 1112.

NABARD (1994); *The Rural Non-Farm Sector in Orissa, Report of the Study Group of the Rural Non-Farm Sector*, National Bank for Agriculture and Rural Development, Bombay.

Nagaraj, R. (1984); "Sub-contracting in Indian Manufacturing Industries: Analysis, Evidence and Issues", *Economic and Political Weekly*, Annual Number, Vol. 19, Nos. 31, 32 and 33, August, pp. 1435-53.

National Small Industries Corporation (undated), *Ancillary Relationship: A Case Study of a Public Sector Unit and its Ancillaries* (mimeo).

Nayak, P. (1993); *Determinants and Consequences of Inter-State Migration from Orissa*, (mimeo), NCDS, Bhubaneswar.

Nayyar, Rohini and Alakh N. Sharma (eds.) (2005); *Rural Transformation in India: The Role of Non-Farm Sector*, Institute for Human Development, New Delhi.

Nelson, Joan M. (1979); *Access to Power: Politics and the Urban Poor in Developing Nations*, Princeton University Press.

Nelson, P. (1959); "Migration, Real Income and Interest", *Journal of Regional Science*, Vol. 1, No. 2, Spring, pp. 43-74.

Ngware, Neena M., S. Suleiman and A. Ngware (2004); "A Case Study of the Issues Faced by Women in the Informal Sector of Urban Tanzania", *Regional Development Dialogue*, Vol. 25, No. 1, Spring, pp. 157-72.

Norcliff, Glen (1983); "Operating Characteristics of Rural Non-Farm Enterprises in Central Province, Kenya", *World Development*, Vol. 11, No. 11, pp. 981-84.

NSSO, *Sarvekshana*, (Various Issues), 27th, 32nd and 38th Rounds, National Sample Survey Organisation, New Delhi.

Nurkse, Ragnar (1966); *Problem of Capital Formation in Underdeveloped Countries*, OUP.

Oberai, A. S. and H. K. M. Singh (1983); *Causes and Consequences of Internal Migration: A Study in the Indian Punjab*, Oxford University Press, Delhi, Chapter-2.

Okafor, F. C. (1983); "Rural Employment Diversification in Anambra State, Nigeria: Its Nature and Implication for Development", *Labour, Capital and Society*, Vol. 16, No. 2, pp. 226-39.

Papola, T. S. (1978); "Informal Sector in an Urban Economy", *Nagarlok*, October-December.

Papola, T. S. (1980); "Informal Sector: Concept and Policy", *Economic and Political Weekly*, Vol. 15, No. 18, May 3, pp. 817-24.

Papola, T. S. (1981a); *Urban Informal Sector in a Developing Economy*, Vikas Publishing House, New Delhi.

Papola, T. S. (1981b); "Dissecting the Informal Sector", *Economic and Political Weekly*, Vol. 16, No. 31, August 1, pp. 1272-74.

Papola, T. S. (1982); *Rural Industrialisation: Approaches and Potential*, Himalaya Publishing House, Bombay.

Papola, T. S. (1987a); "Urbanisation, Rural-Urban Migration and Growth of Informal Sector", in S. Manzoor Alam and F. Alikhan (eds.), *Perspective on Urbanisation and Migration: India and USSR*, Allied Publisher, Ahmedabad.

Papola, T. S. (1987b); "Rural Industrialisation and Agricultural Growth: Case Study of India", in R. Islam (ed.), *Rural Industrialisation and Employment in Asia*, ILO (ARTEP), New Delhi.

Papola, T. S. (1992); "Rural Non-Farm Employment: An Assessment of Recent Trend", *Indian Journal of Labour Economics*, Vol. 35, No. 3, pp. 238-48.

Papola, T. S. (2004); "Globalisation, Employment and Social Protection: Emerging Perspectives for the Indian Workers", *The Indian Journal of Labour Economics*, Vol. 47, No. 3, July-September, pp. 541-50.

Papola, T. S. and V. N. Mishra (1980); "Some Aspects of Rural Industrialisation", *Economic and Political Weekly*, Vol. 15, Nos. 41-43, pp. 1733-46.

Papola, T. S. and R. S. Mathur (1983); "Inter-Sectoral Linkages in the Metal Engineering Industry in Kanpur, India" in S.Watanabe (ed.), *Technology, Marketing and Industrialisation: Linkages between Large and Small Enterprises*. Macmillan, New Delhi.

Peattie, Lisa R. (1982); "What is to be Done with the 'Informal Sector'?: A Case Study of Shoe Manufacturers in Columbia" in Helen I. Safa (ed.), *Towards a Political Economy of Urbanisation in Third World Countries*, OUP, New Delhi.

Portes, Alejandro (1978); "The Informal Sector and the World Economy: Notes on the Structure of Subsidised Labour", *IDS Bulletin*, June, pp. 35-40.

Potts, Deborah (2000); "Urban Unemployment and Migrants in Africa: Evidence from Harare 1985-1994", *Development and Change*, Vol. 31, No. 4, September, pp. 879-910.

Psacharopoulos, George (1988); "Education and Development: A Review", *World Bank Research Observer*, Vol. 3, No. 3, January, pp. 99-116.

Quizon, Jaine and H. P. Binswanger (1986); "Modelling the Impact of Agricultural Growth and Government Policy on Income Distribution", *World Bank Economic Review*, Vol. 1, No. 1, September 1986, pp. 103-48.

Ranis, G. et al. (1990); *Linkages in Developing Economics: A Philippines Study*, ICEG, ICS Press, San Francisco.

Ranis, G. and F. Stewart (1993); "Rural Non-Agricultural Activities in Development: Theory and Application", *Journal of Development Economics*, Vol. 40, pp. 75-101.

Rao, G. Gangadhar (2006); "Variations and Determinants of Rural Non-Farm Employment in AP: An Inter-District Analysis", *The Asian Economic Review*, Vol. 48, No. 1, April, pp. 115-28.

Ray, Sunil (1994); "Farm-Non-Farm Interaction in a Labour Surplus Economy", *Economic and Political Weekly*, Vols. 29, No. 53, December 31, pp. A 171-74.

Rayappa, P. (1986); "Some Dimensions of Off-Farm Employment in Rural Karnataka, India", in R.T. Shand (ed.), *Off-Farm Employment in the Development of Rural Asia*, Vol. 1 and 2, Australian National University, Canberra.

Rempel, Henry (1996); "Rural-to-Urban Migration and Urban Informal Activities", *Regional Development Dialogue*, Vol. 17, No. 1, pp. 37-51.

Remy, Dorothy (1982); "Formal and Informal Sectors of the Zaria, Nigerian Economy: An Analysis Framework with Empirical Content", in Helen I. Safa (ed.) *Towards a Political Economy of Urbanisation in Third World Countries*, OUP, New Delhi, pp. 233-46.

Reynold, L. G. (1969); "Economic Development with Surplus Labour", *Oxford Economic Papers*, Vol. 21, No. 1, March, pp. 89-103.

Reza, Sadrel (1986); "Development of Rural Industry in Bangladesh: Why and How", in Franklin Vivekananda (ed.), *Bangladesh Economy: Some Selected Issues*, Bentham Books, Stockholm.

Richardson, H. W. (1984); "The Role of the Urban Informal Sector: An Overview", *Regional Development Dialogue*, Vol. 5, No. 2, Autumn 1984, pp. 56-62.

Roemer, Michael (1986); "Simple Analytics of Segmented Markets: What Case for Liberalisation?", *World Development*, Vol. 14, No. 3, March, pp. 429-39.

Rogaly, Ben, Danniel Coppard, Abdur Rafique, Kumar Rana, Amrita Sengupta and Jhuma Biswas (2002); "Seasonal Migration and Welfare/Illfare in Eastern India: A Social Analysis", *The Journal of Development Studies*, Vol. 38, No. 5, June, pp. 89-114.

Romatet, Emmanuel (1983); "Calcutta's Informal Sector: Theory and Reality", *Economic and Political Weekly*, Vol. 18, No. 50, December 10, pp. 2115-28.

Romijn, H. (1987); "Employment Generation Through Cottage Industries in Rural Thailand: Potential and Constraint" in R. Islam (ed.), *Rural Industrialisation and Employment in Asia*, ILO/ARTEP, New Delhi.

Rourkela News, various issues.

Saavedra, Jaime and Alberto Chong (1999); "Structural Reform, Institutions and Earnings: Evidence from the Formal and Informal Sectors in Urban Peru", *The Journal of Development Studies*, Vol. 35, No. 4, April, pp. 95-116.

Samaj, The; Cuttack, April 5, 1991.

Samal, Kishor C. (1984); "Migrant Labour—I and II", *Business Standard*, Calcutta, 23rd and 24th June.

Samal, Kishor C. (1986); *Economics of Urban Informal Sector: Case Study of Sambalpur*, Doctoral Dissertation at Giri Institute of Development Studies, Lucknow, for Kanpur University, Kanpur.

Samal, Kishor C. (1990a); *Urban Informal Sector (An Exploration of the Informal Sector in a Small City of Orissa)*, Manak Publication Pvt. Ltd., Delhi.

Samal, Kishor C. (1990b); "Linkages Between Informal and Formal Manufacturing Sector", *Economic and Political Weekly*, Vol. 25, No. 23, June 9, pp. 1287-88.

Samal, Kishor C. (1990c); "Urban Informal Sector and Rural-Urban Migration: A Case Study", *Nagarlok*, Vol. 22, No. 3, July-September, pp. 30-38.

Samal, Kishor C. (1990d); "Urban Informal Manufacturing Sector: Its Structure and Operation (A Case Study)", *Urban India*, Vol. 10, No. 2, December, pp. 55-69.

Samal, Kishor C. (1990e); *Linkages Between Informal and Formal Manufacturing Sectors: Case Study of a Steel City (Rourkela)*, Report of the Research Project funded by ICSSR P.G. Department of Economics, G. M. College, Sambalpur (Orissa).

Samal, Kishor C. (1994a); *Dynamics of Informal Manufacturing Sector: Case Study of Sambalpur at Two Points of Time* (mimeo), Report of the Research Project funded by ICSSR, NCDS, Bhubaneswar.

Samal, Kishor C. (1994b); *Linkages of NALCO Angul Sector with the Manufacturing Sector in Orissa* (mimeo),Project Report for NALCO Chair NCDS, Bhubaneswar.

Samal, Kishor C. (1995); *Socio-Economic Impact of NALCO Angul Sector: Involuntary Displacement and Adjustment* (mimeo), Project Report for NALCO Chair, Nabakrushna Choudhury Centre for Development Studies, Bhubaneswar.

Samal, Kishor C. (1996); *Growth of Informal Activities Around NALCO (NALCO Nagar and Damanjodi Sector)*, (mimeo), Report of the Research Project undertaken for NALCO Chair, NCDS, Bhubaneswar.

Samal, Kishor C. (1997a); "Features and Determinants of Rural Non-Farm Sector in India and Orissa: A Survey", *Journal of Indian School of Political Economy*, Vol. 9, No. 1, January-March, pp. 65-93.

Samal, Kishor C. (1997b); "On Redefining the Urban Informal Manufacturing Sector: A Study of Micro Businesses, Petty Producer Units and Small Capitalist Producer Firms in Sambalpur, Orissa", *The Journal of Entrepreneurship*, Vol. 6, No. 1, January-June, pp. 20-23.

Samal, Kishor C. (1997c); "Informal Sector in an Emerging Urban Industrial Centre and its Policy Implication—Case Study of NALCO Nagar", *Nagarlok*, Vol. 29, No. 3, July-September, pp. 36-49.

Samal, Kishor C. (1997d); "Rural Non-Farm Activities in Specific Regions of Orissa", *Journal of Rural Development*, Vol. 16, No. 3, July-September, pp. 457-64.

Samal, Kishor C. (1998); "Some Aspects of Informal Sector Growth in Orissa", *The Journal of Entrepreneurship*, Vol. 7, No. 1, January-June, pp. 83-91.

Samal, Kishor C. (2000); "Liberalisation and Its Adverse Impact on Informal Sector Workers", *Man and Development*, Vol. 22, No. 2, June, pp. 138-44.

Samal, Kishor C. and Shibalal Meher (1997); *Migration of Workers to Urban Informal Sector (Case Study of Five Class-I Towns in Orissa)*, (mimeo), Report of the Research Project funded by Planning Commission, Government of India, NCDS, Bhubaneswar.

Samal, Kishor C. and Shibalal Meher (1998); "How Geographical Distance Play a Role in Rural-to-Urban Migration and Migrants' Link with their Natives? (A Study of Two Modern, Planned Class-I Towns of Orissa)", *Margin*, Vol. 30, No. 4, July-September. pp. 86-99.

Samal, Kishor C. and Shibalal Meher (2002); "Migrants in Urban Informal Sector: Case Study of Five Class-I Towns and Five Villages in Orissa", *Artha Vijnana*, Vol. 44, No. 1, March, pp. 62-72.

Samal, Kishor C. and Srijit Mishra (1998a); *Migration of People to Talcher Coal Mine Region: A Study of Informal Workers* (mimeo), NCDS, Bhubaneswar.

Samal, Kishor C. and Srijit Mishra (1998b); "Migrant Workers in a Coal Mine Region of Orissa", *The Indian Journal of Labour Economics*, Vol. 41, No. 4, October-December, pp. 745-54.

Samal, Kishor C. and Srijit Mishra (1999); "Determinants of Rural-to-Urban Migration and Migrants in Destinations: A Review of Issues": *Nagarlok*, Vol. 31, No. 1, January-March, pp. 1-12.

Samal, Kishor C., Shibalal Meher and Damodar Jena (1998); "Process of Urbanisation in Orissa: In Search of Explanatory Macro-Economic Factors", *Urban India*, Vol. 18, No. 1, January-June, pp. 85-103.

Sanchez, C. E. et al. (1981); "The Informal and Quasi-Formal Sectors in Cordoba" in S.V. Sethuraman (ed.), *The Urban*

Informal Sector in Developing Countries: Employment, Poverty and Environment, ILO, Geneva, pp. 144-58.

Santos, Milan (1979); *The Shared Space: The Two Circuits of the Urban Economy in Underdeveloped Countries*, Methuen, London and New York.

Sardana, P. K., Veena Manocha and A. C. Gangwar (1995); "Nature and Pattern of Changes in Rural Non-Farm Employment in Haryana", *Indian Journal of Agricultural Economics*, Vol. 50, No. 3, Conference Number, July-September, pp. 417-21.

Sarswat, S. P., Ranveer Singh and Prem Singh Dahiya (1995); "Structural Changes in Farm and Non-Farm Employment in Himachal Pradesh: A Case Study of Kot Village", *Indian Journal of Agricultural Economics*, Vol. 50, No. 3, Conference Number, July-September, pp. 430-36.

Schumacher, E. F. (1973); *Small is Beautiful: A Study of Economics as if People Mattered*, Bland and Brigg, U.K.

Scoville, J. G. (1977); "Sectoral Interdependence in Urban Labour Markets and Variations in the Social and Economic Environment" in Subbaiah Kannappan (ed.), *Studies on Urban Labour Market Behaviour in Developing Areas*, IILS, Geneva, 1977.

Sethuraman, S. V. (1976a); *Jakarta: Urban Development and Employment*, ILO, Geneva.

Sethuraman, S. V. (1976b); "The Urban Informal Sector: Concept, Measurement and Policy", *International Labour Review*, Vol. 114, No. 1, July-August, pp. 69-81.

Sethuraman, S. V. (ed.) (1981); *The Urban Informal Sector in Developing Countries: Employment, Poverty and Environment*, ILO, Geneva.

Shah, Amita (2001); "Scalar Linkages in Industries: Implications for Productivity and Employment" in Amitabh Kundu and A. N. Sharma (eds.), *Informal Sector in India: Perspective and Policies*, IHD and IAMR, New Delhi.

Shand, R. T. (ed.) (1986); *Off-Farm Employment in the Development of Rural Asia*, Vols. 1 and 2, Australian National University, Canberra.

Shaw, Annapurna (1990); "Linkage of Large Scale, Small Scale and Informal Sector Industries: A Study of Thana-Belapur", *Economic and Political Weekly*, Vol. 25, Nos. 7 and 8, February 17-24, M 17-22.

Shukla, Tara (1989); "Rural Non-Agricultural Employment", *Economic and Political Weekly*, Vol. 24, Nos. 35 and 36, September 2-9, pp. 1993-95.

Shukla, Vibhooti (1989); "Rural Non-Farm Activity: A Regional Model and its Empirical Application to Maharashtra", *Economic and Political Weekly*, Vol. 26, No. 45, pp. 1477-88.

Shukla, Vibhooti (1992); "Rural Non-Farm Employment in India: Issues and Policy", *Economic and Political Weekly*, Vol. 27, No. 28, July 11, pp. 1477-88.

Shylendra, H. S. and P. Thomas (1995); "Non-Farm Employment: Nature, Magnitude and Determinants in a Semi-Arid Village of Western India", *Indian Journal of Agricultural Economics*, Vol. 50, No. 3, Conference Number, July-September, pp. 410-16.

Sigurdson, Jon (1983); "The Changing Pattern of Inter-Sectoral Relationship in the Rural Machinery Industry in China", in S. Watanabe (ed.) *op cit*.

Simmons, A. B. (1984); *Population Distribution, Migration and Urbanisation*, Department of International Economics and Social Affairs, United Nations, New York.

Sinclair, S. W. (1978); *Urbanisation and Labour Markets in Developing Countries*, Croom Helm, London.

Singh, Ajit Kumar (1994); "Changes in the Structure of Rural Workforce in Uttar Pradesh: A Temporal and Regional Study", in P. Visaria and R. Basant (eds.), *Non-Agricultural Employment in India: Trends and Prospective*, Sage Publications, New Delhi.

Sjaastad, L. A. (1962); "The Costs and Returns of Human Migration", *Journal of Political Economy*, Vol. 70, No. 5, pp. 80-93.

Smith, Charles David (2001); *Ecology, Civil Society and the Informal Economy in North West Tanzania*, Aldershot, Ashgate.

Souza, Paulo R. and Victor E. Tokman (1976); "The Informal Urban Sector in Latin America", *International Labour Review*, Vol. 114, No. 3, November-December, pp. 355-65.

Sreeramamurty, Kudamala (1986); *Urban Labour in Informal Sector*, B. R. Publishing Corporation, Delhi.

Stark, Oded (1982a); "Research on Rural-to-Urban Migration in LDCs: The Confusion Frontier and Why We should Pause to Rethink Afresh", *World Development*, Vol. 10, No. 1, pp. 63-70.

Stark, Oded (1982b); "On Modelling the Informal Sector", *World Development*, Vol. 10, No. 5, pp. 413-16.

Steel Authority of India Ltd.; (undated), *Rourkela Steel Plant: Annual Implementation Plan for Modernisation and Major A.M.R. Schemes*, 1990-91, Raurkela.

Steel, W. F. (1976); "Economic Measurement of the Relative Size and Productivity of Intermediate Sector Employment: Some Estimates from Ghana", *Manpower and Unemployment Research*, Vol. 9, No. 1, April, pp. 23-31.

Steel, W. F. (1997); *Small-Scale Employment and Production in Developing Countries: Evidence From Ghana*, Praeger Publisher, New York.

Suryanarayan, S. S. (1998); "Information Base for the Informal Sector", *The Indian Journal of Labour Economics*, Vol. 41, No. 3, July-September, pp. 473.

Thorat, Sukhdeo (1993); "Land Ownership Structure and Non-Farm Employment of Rural Household in India", *Indian Journal of Labour Economics*, Vol. 36, No. 4, July-September, pp. 470-91.

Todaro, M. P. (1976); *Internal Migration in Developing Countries*, ILO, Geneva.

Todaro, M. P. (1969); "A Model of Labour Migration and Urban Unemployment in Less Developed Countries", *American Economic Review*, Vol. 59, No. 1, March, pp. 138-48.

Todaro, M. P. (1971); "Income Expectations, Rural-Urban Migration and Employment in Africa", *International Labour Review*, November, pp. 391-95.

Tokman, Victor E. (1978a); "An Exploration into the Nature of Informal-Formal Sector Relationship", *World Development*, Vol. 6, No. 9/10, September/October, pp. 1065-75.

Tokman, Victor E. (1978b); "Competition Between the Informal and Formal Sector in Retailing: The Case of Santiago", *World Development*, Vol. 6, No. 9/10, September/October, pp. 1187-98.

Unni, Jeemol (1989); "Changes in Women's Employment in Rural Areas 1961-1983", *Economic and Political Weekly*, Vol. 24, No. 17, April 29, pp. WS-6 WS-9

Unni, Jeemol (1991); "Regional Variation in Rural Non-Agricultural Employment: An Exploratory Analysis", *Economic and Political Weekly*, Vol. 26, No. 3, January 19, pp. 109-22.

Unni, Jeemol (1994); "Inter-regional Variations in Non-Agricultural Employment in Rural India: An Exploratory Analysis", in P.Visaria and R. Basant (eds.), *Non-Agricultural Employment in India: Trends and Prospects*, Sage Publications, New Delhi.

Unni, Jeemol (2005); "Wages and Incomes in Formal and Informal Sectors in India", *The Indian Journal of Labour Economics*, Vol. 48, No. 2, April-June, pp. 311-17.

Vaidyanathan, A. (1986); "Labour Use in Rural India: A Study of Spatial and Temporal Variation", *Economic and Political Weekly*, Vol. 21, No. 52, December 27, pp. A130-A146.

Vaidyanathan, A. (1994); "Employment Situation: Some Emerging Perspectives", *Economic and Political Weekly*, Vol. 29, No. 50, December 10, pp. 3147-56.

Verma, Binoy N. and Neelam Verma (1995); "Distress Diversification from Farm to Non-Farm Rural Employment Sector in the Eastern Region", *Indian Journal of Agricultural Economics*, Vol. 50, No. 3, Conference Number, July-September, pp. 422-29.

Visaria, Pravin (1995); "Rural Non-Farm Employment in India: Trends and Issues for Research", *Indian Journal of Agricultural Economics*, Vol. 50, No. 3, Conference Number, July-September, pp. 398-409.

Visaria, Pravin and Rakesh Basant (eds.) (1994); *Non-Agricultural Employment in India: Trends and Prospective*, Sage Publication India Pvt. Ltd.

Vogel, S. (1994); "Structural Changes in Agriculture: Production Linkages and Agricultural Demand-Led Industrialisation", *Oxford Economic Paper*, Vol. 46.

Waldorf, Saral Teilhat and William H. Waldorf (1983). "Earnings of Self-employed in an Informal Sector: A Case Study of Bangkok", *Economic Development and Cultural Change*, Vol. 31, No. 3, April, pp. 587-607.

Walle, Van don D. (1994); *Rural Poverty in an Emerging Market Economy: Its Diversification into Non-Farm Activities in Rural Vietnam: The Solution*, Policy Research Department, World Bank, Washington, (mimeo.), October.

Watanabe, Susumu (ed.) (1983a); *Technology, Marketing and Industrialisation: Linkages Between Large and Small Scale Enterprises*, Macmillan, New Delhi.

Watanabe, Susumu (1983b); "Technical Cooperation Between Large and Small Firms in the Filipinos Automobiles Industry" in S. Watanabe (ed.), *opcit*, pp. 97-132.

Watanabe, Susumu (1983c); "Technological Linkages Through Sub-Contracting in Mexican Industries" in S. Watanabe (ed.) *op.cit*, pp. 169-92.

Weeks, John (1975); "Policies for Expanding Employment in the Informal Urban Sector of Developing Economies", *International Labour Review*, Vol. 111, No. 1, January, pp. 1-13.

Weeks, John (1977); "An Exploration into the Nature of the Problem of Urban Imbalance in Africa", *Manpower and Unemployment Research in Africa*, Vol. 6, No. 2, November, pp. 9-36.

White, B. (1991); "Economic Diversification and Agrarian Change in Rural Java 1900-1990", in P. Alexander, et al. (eds.), *In the Shadow of Agriculture: Non-Farm Activities in the Javanese Economy: Past and Present*, Royal Tropical Institute, Amsterdam.

Williamson, J. G. (1988); "Migration and Urbanisation", in H. Chennery and T. N. Srinivasan (eds.), *Handbook of Development Economics*, Vol. 1, Amsterdam, North-Holland.

World Bank (1978); *Rural Enterprises and Non-Farm Employment*, World Bank, Washington.

World Bank (1995); *World Development Report 1995: Workers in an Integrating World*, OUP, New York.

Yap, L. (1977); "The Attraction of Cities: A Review of the Migration Literature", *Journal of Development Economics*, Vol. 4, No. 4, pp. 239-326.

Index